XVIII Brazilian Colloquium on Orbital Dynamics (CBDO 2016)

AA005599

Journal of Physics: Conference Series Volume 911

Aguas de Lindoia, Brazil
28 November - 2 December 2016

Editors:

Cristiano F. de Melo
Antonio F. Bertachini de Almeida Prado
Elbert E. N. Macau

ISBN: 978-1-5108-5039-2
ISSN: 1742-6588

Printed from e-media with permission by:

Curran Associates, Inc.
57 Morehouse Lane
Red Hook, NY 12571

Some format issues inherent in the e-media version may also appear in this print version.

Printed by Curran Associates, Inc. (2017)

For permission requests, please contact the Institute of Physics
at the address below.

Institute of Physics
Dirac House, Temple Back
Bristol BS1 6BE UK

Phone: 44 1 17 929 7481
Fax: 44 1 17 920 0979

techtracking@iop.org

Additional copies of this publication are available from:

Curran Associates, Inc.
57 Morehouse Lane
Red Hook, NY 12571 USA
Phone: 845-758-0400
Fax: 845-758-2633
Email: curran@proceedings.com
Web: www.proceedings.com

Table of contents

Volume 911

XVIII Brazilian Colloquium on Orbital Dynamics
28 November to 2 December 2016, Águas de Lindoia (SP), Brazil

Accepted papers received: 12 October 2017
Published online: 25 October 2017

Preface

Papers

XVIII Brazilian Colloquium on Orbital Dynamics (2016): the bases of Celestial Mechanics and its development in the research institutions in Brazil

Cristiano F. de Melo[1], Antônio F. Bertachini de Almeida Prado[2], Elbert E. N. Macau[2,3]

[1] Federal University of Minas Gerais, UFMG, Belo Horizonte (MG), Brazil
[2] National Institute for Space Research, INPE, São José dos Campos (SP), Brazil
[3] Federal University of São Paulo, UNIFESP, São Jose dos Campos (SP), Brasil

E-mail: cristiano.fiorilo@demec.ufmg.br

Abstract. The Brazilian Colloquium on Orbital Dynamics (CBDO) is the most important and traditional scientific meeting aimed at Theoretical and Applied Celestial Mechanics in Latin America context, and it is realized every two years. The last one, XVIII CBDO edition, took place from November 28 to December 02, 2016, in the city of Águas de Lindoia (SP) Brazil, and received more than two hundred researchers from Brazil, Argentina, Uruguay, Peru, Mexico, United States, Europe and Asia. For five days, current and high-level works in Celestial Mechanics were presented and discussed. This special issue of the Journal of Physics: Conference Series brings together a set of 27 papers presented in the CBDO distributed in the areas covered by the meeting.

1. Introduction

Celestial Mechanics is an area of Astronomy that studies the dynamics of the bodies under the action of gravitational interaction, including both spacecrafts and natural astronomical bodies such as star, planets, moons, comets, planetary rings, etc.

The beginnings of the Astronomy and Celestial Mechanics is related with the observation and description of the periodicities in the celestial bodies' motion since the early days of mankind. As such, ancient astronomers learned the differences between stars and planets, as stars remain relatively fixed over the centuries while the planets have movement in comparatively short time, as well as many other phenomena. As a result, this science played an important role in determining the seasons, calendarizes and in the development of agriculture, navigation and religion, thus, contributing to the emergence of important ancient civilizations and their cultures [1].

Modern Celestial Mechanics started with the works of Nicolaus Copernicus (1473-1543), Tycho Brahe (1546-1601), Johannes Kepler (1572-1630), Galileo Galilei (1564-1642) and Isaac Newton (1642-1727). Kepler, for example, provided important contributions as a model for quantitative prediction of planetary position from geometrical and arithmetical techniques, and he was also the first to rise discussions on the physical causes of the motion of the planets [2]. In this historical context, Newton presented the first formulation of gravitational force, which was published in Phylosophiae Naturalis Principia Mathematica, in 1687, describing the causes of the planetary motion and expanding the Kepler's laws. However, the term "Celestial Mechanics" is more recent than that. It was introduced by Pierre-Simon Laplace (1749 - 1827) [2]. During the centuries XVIII and IXX, many other scientists contributed to the bases and advances in Celestial Mechanics and for the understanding of the origin and evolution of the Solar System. Among them, we can cite Leonhard Euler (1707 - 1783), Jean le Rond d'Alambert (1717 - 1783), Giuseppe Lagrange (1736 - 1813), the Laplace himself, Adrien-Marie Legendre (1752 - 1833), Hamilton William (1805 - 1865), Lyapunov Aleksandr (1857 - 1918), and many others. But this list of honorable scientists would not be complete without Henri Poincaré (1854 - 1912), the most important contributor to Celestial Mechanics in the final years of the nineteenth and the early years of the twentieth centuries. Considering the circular restricted three-body problem, where two massive bodies move in circular orbits around their common center of mass, and a particle with negligible mass moves under the gravitational attraction of the other two, Poincaré established the concepts of nonintegrable dynamical system and deterministic chaos [3]. Although the circular restricted three-body problem is described by a set of deterministic equations, its behavior may become chaotic and unpredictable under certain conditions. However, taking account the mass hierarchy in the solar system, the work of Poincaré can be considered to describe many sub systems such as Sun-planet-particle, Earth-Moon-particle, etc. [3, 4], as well as countless strategies to space navigation in the context of modern Astronautics.

The works of these scientists form the bases of several current researches in theoretical and applied Celestial Mechanics, including the ones presented in this special issue.

2. Space age

In the twentieth century, after the end of Second World War, the space age began, firstly driven by a frenzied race between United States of America and the former Soviet Union. The milestone of this age was the launch of the Sputnik satellite [5]. So, a new and exciting branch of the Celestial Mechanics so-called Orbital Mechanics emerged with focus on the study on rocket and spacecraft trajectories, including orbital maneuvers such as orbital plane changes, lunar and interplanetary transfers as well as studies on propulsion systems and devices to control these space vehicles.

Thus, from the 50's years the mankind began its adventure toward space, and soon in these first 60 years of this new age, powerful rockets have taken thousands of artificial satellites and space stations into space. Astronauts have walked on the Moon, more than 100 automatic probes have launched toward moon, all planets of the solar system and many asteroids and comets [5]; the space telescopes Hubble and Kepler are right now in orbit exploring the deep space and finding exoplanets and new planetary systems, thereby, providing new knowledges and challenges for Celestial Mechanics [1], and helping us to understanding better our place and mission in the Universe.

This demand for space navigation is clearly responsible for the recent progress in Celestial Mechanics, and it is strongly linked with the advent of high-performance digital computers. Indeed, the needed accuracy for spacecraft navigation would not be possible without the advances of the high-speed computers. For examples, to put the Pionner II probe (launched in April 1973) in a trajectory that reached Jupiter in December 1977 and Saturn in September 1979, it was required a navigation accuracy of better than one part in 10 million [3, 5]. Much more precision and intensive calculations were required on the Cassini-Huygens probe which studies Saturn and its natural satellites since 2004, including a landing on Titan and testing the Einstein's theory of relativity [6]. Another fantastic adventure through the space was the Rosetta mission. Its journey started on 2 March 2004. The spacecraft performed fly-bys with the Earth (November 2005, 2007 and 2009) and Mars (February 2007), the asteroids 2867 Steins and 21 Lutetia on September 2008 and July 2010, respectively, before of reaching its target – the comet 67P/Churyumov-Gerasimenko – on August 2014, and becoming the first spacecraft to follow a comet. The Philae module was released by the spacecraft on December 12, 2014, and, in turn, it was the first human device to land on a comet's surface [7]. Missions like these are only possible thanks to the accuracy, robustness and durability of the currently high-performance computers. Many other examples can be given, but exceed the purposes of this editorial.

Currently, the Orbital Mechanics has a very important role for modern society, since there are thousands of satellites in Earth's orbit providing fast communication, entertainment and navigation facilities, weather forecast and natural resource prospecting. If today we live the reality where large industrial or trade corporations, wars and homemade routines can work in a continent and to be managed from other ones; and whether it is possible to be with friends and relatives without known their exact location around the world, such as predicted by Sir Arthur Clarke in the early 1960's, it is because the advances of the space exploration and the fantastic machines that are in space. However, the man still didn't learn to live without pollutes and, for this reason, the space activities also spread debris around the Earth that threatening the space exploration. How to predict the motion of those debris and how to remove them are nowadays important topics in actual Orbital Mechanics.

Many challenges for the future of space exploration propel the current researches in Orbital Mechanics, for example, the control over the space debris, the resumption of space flights to the Moon and the dreamt colonization of the Solar System, and, why not, other systems. How? If we can evoke the romantic image of the majestic sailing ships that crossed the ocean with goods and settlers to the New World between the 16th and 19th centuries, we can also imagine big spaceships propelled by Solar sails cruising the depths of interplanetary space transporting goods and settlers to truly new worlds.

3. Orbital Mechanics in Brazil

Activities related to space area in Brazil began in the mid-1950s. At that time, the Brazilian Ministry of Aeronautics worked to build and entrench the Technical Center of Aeronautics (CTA – Centro Técnico da Aeronáutica) located in the city of São José dos Campos (SP). Currently, it is called Department of Aerospace Science and Technology (DCTA - Departamento de Ciência e Tecnologia Aeroespacial), and it is subordinated to the Brazilian Ministry of Defense [5,8]. DCTA has several institutes, among which we highlight:

- Institute of Aeronautics and Space (IAE – Instituto de Aeronáutica e Espaço), created in 1969 [5,9], is the main responsible for the development, production and launching of rockets in Brazil. During the course of its 37 years, it has developed the sounding rockets: Sonda I, II, III and IV, VS-30, VS-40, VSB30 and the VLS (satellite launch vehicle), among other projects [10].
- Technological Institute of Aeronautics (ITA – Instituto Tecnológico da Aeronáutica), created in 1950 [5], is aimed to the teaching and the research in Aeronautics, Mechanics, Electronic, Computation, Airport infrastructure and Aerospace Engineering. ITA had crucial importance in the development of the Brazilian aerospace industry, such as EMBRAER, for example, and also in Orbital Mechanics.

From 1960, the Brazilian government created several groups and committees to leverage researches in the space area. In 1971, these groups and committees gave origin to the National Institute for Space Research (INPE – Instituto Nacional de Pesquisas Espaciais), also located in the city of São José dos Campos (SP), and with the mission of conducting scientific researches, technological development and to provide qualified human resources in the space areas. Currently, INPE has sites located in 11 cities in the Brazilian territory [5,12], and it is subordinated to Brazilian Ministry of Science, Technology, Innovation and Communication.

INPE has already developed and launched 7 satellites, two of which are satellites for data collection (SCD 1 and 2) and 5 others for terrestrial resources in cooperation with China (CEBERS project). Currently, programs as the satellite Amazon 1 for Earth observation and of small scientific satellites as EQUARS for atmosphere monitoring and MIRAX for investigation of the galactic plan through X-rays, and a Multi Mission Platform for satellites are examples of projects in development in INPE [13].

The Institute has several divisions and laboratories to carry out its mission. In the field of applied Celestial Mechanics, there are the Space Mechanics and Control Division, the Laboratory for Integration and Testing of Satellites, the Combustion and Propulsion Division, as well as the postgraduate courses in the areas of Space Engineering and Technology with emphasis in Space Mechanics and Control, Combustion and Propulsion, Science and Technology of Materials and Sensors, and Space Systems Engineering and Management. INPE also offers postgraduate courses in other areas such as Astrophysics, Earth System Sciences, Space Geophysics, Meteorology and Remote Sensing and Applied Computing.

Teaching and research in Orbital Mechanics has made great advances since the foundation of the ITA and INPE. The training of human resources by these institutes is undoubtedly of great importance for the diffusion of researches in Orbital Mechanics in Brazil. Researchers trained in INPE and ITA Postgraduate courses hold positions as professors and researchers in several public and private universities in the country, for example, in the Federal Universities of ABC (UFABC), São Paulo (UNIFESP), Paraná (UFPR), Brasília (UNB), Minas Gerais (UFMG), Recôncavo da Bahia (UFRBA), São Paulo State University (UNESP), Feira de Santana State University – Bahia (UEFS), State University of Campinas (UNICAMP), Pontificial Chatolic University (PUC), etc. In this way, cooperation networks between these universities, the INPE and the ITA arise and are strengthened day-by-day. This special issue of the Journal of Physics: Conference Series is a good example of this mechanism.

Theoretical Celestial Mechanics, in turns, had its development in Brazil linked to the Institute of Astronomy, Geophysics and Atmospheric Sciences (IAG) of the University of São Paulo (USP) [14], National Observatory (ON) [15] and Federal University of Rio de Janeiro (UFRJ). Similarly, postgraduate courses at these institutions have formed many Professors and Researchers who work in several public and private universities in Brazil.

4. Brazilian Colloquium on Orbital Dynamics, CBDO

In the early 1980's, a small group of Brazilian scientists organized the first edition of the Brazilian Colloquium on Orbital Dynamics [1]. Since then, the CBDO has taken place every two years, gathering scientists from South America as well as guests from other continents. The main areas of the CBDO deals with are: theoretical Celestial Mechanics, Dynamical and Planetary Astronomy, Planetary Science and Solar System dynamics, Astronautics and Dynamical Systems and Chaos applied to space research.

The 27 papers selected for this special issue have between six and nine pages and bring together the outcomes of research studies developed in Brazilian universities and research institutes. They represent the results of the strengthening of important research groups and the formation of new researchers and engineers who will work in Brazilian space activities in the coming decades. In the next section, the papers are presented one by one.

5. Introducing the papers

In this section, we present the 27 selected papers of this special issue. We also take this opportunity to thank all the authors for their valuable contributions, and for investing a lot of work and effort into the results of their research works. Special thanks go to reviewers who dedicated their precious time in providing numerous comments and suggestions, criticism and constant and enthusiastic support. The papers are presented one by one in sequence.

The paper "Searching for orbits around the triple system 45 Eugenia" searches for orbits to place a spacecraft around the triple asteroid 45 Eugenia [16]. It uses a semi-analytical mathematical model that represents the most important dynamical characteristics of that system [17]. The smaller bodies are assumed to be in elliptical orbits that are processing due to the flattening of the main body. Orbits with passages near both smaller bodies are considered. The simulations consider other dynamical aspects, such as close approaches [18] and gravitational captures [19].

The paper "Fractal boundaries in chaotic Hamiltonian systems" considers the motion of charge particles in a magnetized plasma subjected to electrostatic drift waves and identify the set of initial conditions that imply in trajectories that leave the interaction region through one of two different and complementary exits. They show that the initial conditions associated to those exists belongs to two sets that present a fractal basin boundary [20]. This behavior leads to a dynamics that phenomenon of final-state uncertainty [21], i.e., once the initial condition is determined within a certain uncertainty, it turns to be increasingly difficult to know for sure to which exit the trajectory will pass through.

The paper “Deviations in CBERS-4 Satellite Direction Components From The Electromagnetic Disturbance of Communication Antennas" brings a model for disturbance of electromagnetic origin [22,23] on the orbit of the CBERS-4 (China-Brazil Earth Resources Satellite 4) considering the antennas theory and the laws of the energy-momentum conservation [24].

The paper "Gamma radiation in ceramic capacitors: a study for space missions" investigates the effects of the cosmic radiation on ceramic capacitors [25]. The investigation uses a data logging system developed considering the open source Arduino platform and shows how the capacitance decreases with exposure to gamma radiation [26].

The effects on a spacecraft's electrical equipments during its passage through the Van Allen belts [27] are investigated in the paper "Analysis of the passage of a spacecraft between the Van Allen belts considering a low and high solar activity". The Van Allen belts are modeled using the data from the Van Allen probes mission to estimate the spacecraft's absorbed radiation dose [28,29].

A solar energy collector system that can operates 24 hours per day is the subject of the paper entitled “Solar Power Satellite system in formation on a common geostationary orbit”. The authors propose and analyze a system composed by a sunlight reflector and a microwave transmitting satellite (MTS) orbiting the Earth. The microwave transmitting satellite (MTS) is placed on a common geostationary orbit (GEO) in the Earth's equatorial plane, and the sunlight reuses the solar radiation pressure to achieve quasi-periodic orbits about the MTS, so that the sunlight is always redirected to the MTS, which converts the solar energy in electromagnetic power and transmits it by microwaves to an Earth-receiving antenna. A linear approximation found by Baig and McInnes [30] was used as an initial condition for the full equations of motion of the sunlight reflector about the GEO point in the Earth-satellite two-body problem, taking into account the effects solar radiation pressure.

The paper “Searching for orbits around the triple asteroid 2001SN263” searches for orbits for a spacecraft that needs to observe the smaller bodies of this triple asteroid system [31,32]. The orbits are classified using the criterion of minimum average distance between the spacecraft and the asteroid to be observed. Stable orbits around both smaller bodies were found with average distances below 1.5 km. The mathematical model takes into account the gravity of all the bodies and the solar radiation pressure.

The paper “On the use of a variable coefficient of reflectivity associated with an augmented area-to-mass ratio to de-orbit CubeSats” shows a new option to de-orbit small satellites from low Earth orbits. It uses a device for area augmentation and/or variation of the coefficient of reflectivity of the satellite [33-35]. Using this technique makes to de-orbit the satellite even in regions where the density of the atmosphere is small or almost zero not.

The evaluation of disturbance of the four Galilean moons (Io, Europa, Ganymede and Callisto) on a spacecraft’s trajectory orbiting Jupiter was the aim of the paper “Mapping the Galilean moon’s disturbance acting on a spacecraft’s trajectory”. The focus is on a short period of time, a few days, equivalent to the orbital period of the farthest Galilean moon from Jupiter, to evaluate the magnitude of the Galilean moons gravitational disturbance applied in the spacecraft trajectory during this period of time. The authors were able to show the disturbance velocity increments in the spacecraft trajectory caused by that four Galilean moons. The simulations were performed using the STRS (Spacecraft Trajectory Simulator), developed by [36, 37].

The dynamics of artificial satellites in Mercury's orbits when the effects due non-spherical shape of this planet [38] is considered and discussed in the paper "Analysis of the long and short-period terms due the nonsphericity of the central body: applications for Mercury". The investigation consists in an approach taking into account the single-averaged and unaveraged equations. Emphasis is given to analyze the effect of the C_{22} term in the dynamics of the spacecraft in special in the variation of the spacecraft orbital elements [39,40].

The paper “Injection of a microsatellite in circular orbits using a three stage launch vehicle” develops a study to verify the capacity of the Brazilian Microsatellite Launch Vehicle (VLM) to inject payloads into Low Earth Orbits [41, 42]. The dynamical model used in the paper has three degrees of freedom associated with the translational motion of the rocket. It is shown the altitude reached in the separation of the second stage, the altitude and velocity of injection, the flight path angle at the moment of the activation of the third stage and the duration of the ballistic flight. Numerical integrations are used to approach this problem [43].

A tetrahedral layout of four satellites in flying formation is presented in paper "Relative Positioning Evaluation of a Tetrahedral Flight Formation's Satellites". The formation is calculated considering a geometrical perspective and the adjust the orbital parameters of each satellite. Also, A detection algorithm is used as flag to signal the regular tetrahedron's exact moments of occurrence. To do so, the volume calculated during the simulation is compared to the real volume, based on the initial conditions of the exact moment of formation and respecting a tolerance [44].

A study on the application of the Lambert's problem [45] of bi-impulsive orbital transfers with time constraint for the execution of minimum fuel maneuvers between coplanar and non-coplanar elliptical orbits is presented in paper "Study of orbital transfers with time constraint and fuel optimization". A genetic search algorithm [46] is considered to determine the minimum fuel trajectory with an iterative variation of two Keplerian elements of the final orbit [47].

In the paper "Orbital and attitude evolution of SCD-1 and SCD-2 Brazilian satellites", the authors present the time history of the SCD satellites orbit and attitude, and to present some discussions about the evolution of these parameters. The SCD-1 and SCD-2 satellites were launched in 1993 and 1998, respectively. For more than 20 and 16 years in service, they have shown reliable operation. The main objective of this work is to present the time history of the SCD satellites orbit and attitude, and to present some discussions about the evolution of these parameters. The rate of change of their orbital elements is small and the main dissipative orbit perturbations acting on the satellites are the sun radiation pressure and atmospheric drag in a less extent. The authors using estimation obtained by simulations show that the satellites will reach the highest decay around the year 2030. The satellites reentry is estimated to occur well after the year 2120. Considering that the maximum operational time was designed, at the launching time, to be around 2 years for each satellite, the conclusion is that the excellent Brazilian manufacturing and engineering made it possible for the satellites to be still operating and collecting data, and to the date there are no signs of any fatal malfunction which could decommission the satellites.

The laser rangefinder to be used in the ASTER deep space mission is the main subject of the paper entitled "Reviewed plan of the ALR, the laser rangefinder for the ASTER deep space mission to the triple asteroid 2001-SN263". The Brazilian deep space mission ASTER plans to send a small spacecraft to investigate the triple asteroid 2001-SN263. The launch is expected to occur in June, 2020. In this work, the most important features for the laser rangefinder are defined and presented.

Studies on clouds of debris close a spacecraft [48,49] and test of maneuvers for propulsion systems with linear and exponential mass variation [50] is the subject of the paper "Collisional Cloud Debris and Propelled Evasive Maneuvers". These studies show that the propulsion system with linear mass variation model is more efficient.

The paper "Artificial satellites orbiting planetary satellites: critical inclination and sun-synchronous orbits" study the behavior of the critical inclination and sun-synchronous orbits for artificial satellites that are orbiting moons of planets. The study takes into account the effects of the harmonics J_2 and C_{22}. The moons considered in the simulations are the Moon [51-53] of the Earth, Io and Europa [54].

The asteroid 243 Ida is located in the asteroid belt, between Mars and Jupiter, belongs to the family of asteroids Koronis whose origin is supposed to be the result of the collision of two major bodies [55]. In the paper "Simulation of the trajectories described by a space vehicle around the asteroid 243 Ida and its natural satellite Dactyl" the authors study the dynamics of this system, orbital trajectories are simulated around Ida considering, besides the gravitational attraction of Dactyl, the non-central gravitational field of the asteroid, defined by a polyhedral model that defines the shape and the non-uniform mass distribution of the body. The simulations showed that due to the irregular shape of the asteroid an exploration spacecraft would approach the surface even when using an almost circular initial orbit. This is a characteristic of the orbits around asteroids, in which the gravitational perturbation can oscillates expressively as function of the altitude. The main disturbance in the trajectory of the spacecraft was due to the gravitational potential of Ida.

The paper "A 2-DOF model of an elastic rocket structure excited by a follower force" brings a model of an elastic rocket structure excited by the follower force given by the motor thrust tangent to the deformed shape [56,57]. This model has two degree of freedom and the results indicate possible occurrence of stable and unstable vibrations, such as limit cycles [58].

Electromagnetical propulsion is the main subject of the paper "Particle-in-cell numerical simulations of a cylindrical Hall thruster with permanent magnets". Hall thrusters (HTs) are electromagnetic propulsion devices that apply a cross-field discharge in an annular channel to generate a plasma [59]. The cylindrical Hall thruster (CHT) is an alternative concept first developed at the Princeton Plasma Physics Laboratory [60]. The CHT presents higher propellant utilization and comparable performance with conventional HTs [61]. The magnetic field in this thruster can be set by using electromagnets or permanent magnets [62]. In this paper, the authors present first results of a numerical model of a CHT. This model solves particle and field dynamics self-consistently using a particle-in-cell approach. A number of techniques applied to reduce the execution time of the numerical simulations is also described. For this, a strategy is used in which a spacecraft orbits Mars, but with the same orbital period of Deimos and similar orbital elements.

In the paper "Orbital Trajectories in Deimos Vicinity Considering Perturbations of Gravitational Origin", the authors study the possibility of maintaining an artificial satellite near to the Deimos surface, even when considering the intense Mars gravitational attraction. This approach of the spacecraft to Deimos, as well as to maintain the spacecraft in this region could be of great importance for observation, conducting experiments or even landing.

The paper "Equilibrium points in the asteroid 2001SN263" searches for the equilibrium points of the triple asteroid system 2001SN263. It takes into account the sizes and shapes of the bodies that constitute the system [63-66]. The study is made using the physical and orbital characteristics of the largest body. Nine equilibrium points are found, with five of them collinear. An analysis of the stability these points are also made, showing that only the equilibrium points L4 and L5 are linearly stable. The results can be used for the ASTER mission under planning [67].

Electromagnetical propulsion is also the main subject of the paper "Magnetic Field Design for a Strongly Improved PHALL Thruster". Here, the focus is on the developing high efficiency Hall Effect Plasma Thrusters (PHALL) using permanent magnets [68]. The use of permanent magnets in these thrusters is mainly related to the decrease of used power for propulsion, especially important for low power thrusters as for micro-satellites. The authors show how each magnetic field configuration affects the generated plasma and consequently the generated propulsion force and efficiency.

The paper "Rendezvous and quasi-rendezvous maneuvers with space debris" studies the problem of rendezvous maneuvers between a spacecraft and a space debris [69-73]. This technique can be used to help to remove space debris and so to protect the Earth and other satellites against collisions. It is studied the distribution of these maneuvers in terms of the final relative velocities between the spacecraft and the debris. The results show that it is possible to use these maneuvers to approach the debris.

An investigation on the application of a flat Solar sail as method of propulsion [74-76] is considered in paper "Numerical analysis of orbital transfers to Mars using solar sails and attitude control". A attitude control that allows a finite variation number in the orientation of the sailcraft along of its trajectory guarantees reduction of the final relative velocity with Mars without, however, a very larger increase in time of transfer.

The paper "Orbital migration and Resonance Offset of the Kepler-25 and K2-24 systems" presents analytical and numerical studies of the resonance capture under Type-I migration for the Kepler-25 [77] and K2-24 [78] Kepler systems. Those two systems are close to a 2/1 mean-motion resonance. The results show that the average values of the period-ratio between two consecutive planets have an important shift compared to the resonant nominal value. The results depend on the flare index and the distance to the central star. The results obtained are in agreement with the available observations.

References

[1] de Melo C F *et al* 2015 Celestial Mechanics: from the bases of the past to the challenges of the future *J. Phys.: Conf. Series* **641**
[2] Morão R R F 1995 *Dicionário Enciclopédico de Astronomia e Astronáutica* 2ª Ed (Rio de Janeiro: Editora Nova Fronteira)
[3] Szebehely V G, Mark H 1998 *Adventures in Celestial Mechanics* 2nd edition John Wiley & Sons, Inc
[4] Murray C D, Dermott S F 1999 *Solar Sytem Dynamics* (Cambridge University Press)
[5] Winter O C, Prado A F B A 2007 *A Conquista do Espaço, do Sputnik à Missão Centenário* (São Paulo, Editora: Livraria da Física)
[6] https://saturn.jpl.nasa.gov/, acceded on September 01, 2017
[7] http://www.esa.int/Our_Activities/Space_Science/Rosetta, accessed on September 09, 2017
[8] http://www.cta.br/index.php/historico, accessed on September 24, 2017
[9] http://www.iae.cta.br/, accessed on September 4, 2017
[10] http://www.iae.cta.br/index.php/todos-os-projetos, accessed on September 242, 2017
[11] http://www.ita.br/grad, accessed on September 24, 2017
[12] http://www.inpe.br/institucional/sobre_inpe/instalacoes.php, accessed on September 25, 2017
[13] http://www.ete.inpe.br/programas.php, accessed on September 25, 2017
[14] http://www.astro.iag.usp.br/~dinamica/cbdo.htm, accessed on September 25, 2017
[15] http://www.on.br/index.php/pt-br/, accessed on September 25, 2017
[16] Marchis F *et al* 2010 A dynamical solution of the triple asteroid system (45) Eugenia *Icarus* **210** pp 635-643
[17] Masago B Y P L *et al* 2016 Developing the ''Precessing Inclined Bi-Elliptical Four-Body Problem with Radiation Pressure" to search for orbits in the triple asteroid 2001SN263 *Adv. Space Res.* 5796
[18] Gomes V M, Formiga J K S, Moraes R V 2013 Studying Close Approaches for a Cloud of Particles Considering Atmospheric Drag *Mathematical Problems in Engineering* (Print) **2013** pp 1-10
[19] Machuy A L, Prado A F B A, Stuchi T J 2007 Numerical study of the time required for the gravitational capture in the bi-circular four-body problem *Adv. Space Res.* **40**(1) pp 118-124
[20] Bleher S, Grebogi C, Ott E, Brown R 1988 Fractal boundaries for exit in Hamiltonian dynamics *Phys. Rev.* A **38** pp 930-938
[21] McDonald S W, Grebogi C, Ott E, Yorke J A 1985 Fractal basin bound-aries *Phys. Rev.* D **17** pp 125-153
[22] Bukhsh W A, Jonsson B L G 2010 Element Position Perturbation for a Narrow Spot Beam with Applications to Satellite Communication Antennas *Progress in Electromagnetics Research* PIER **104**, pp 283- 295
[23] Heilmann A *et al* 2013 Antenna Radiation Effects on the Orbits of GPS and INTELSAT satellites *Acta Astronautica* **88** pp 1–7
[24] Zhou B, Liu Q, He, X 2008 Research on the Electromagnetic Interference of Antennas on the Satellite *Piers online* **4** 3
[25] Ferreira E S, Gonçalves K A, Mittani J C R, Yee M and Tatumi S H 2014 Gamma radiation effects in capacitance of alpha alumina doped with rare earths Radiation physics and chemistry **95** 385
[26] Gonçalves K A *et al* 2010 Study of radiation induced effects in the luminescence of nanostructured Al2O3: Y b *Er crystals IOP Conf. Ser. Mater. Sci. Eng.* **15** 012080
[27] Van Allen J A, Ludwig G H, Ray E C and Mcilwain C E 1958 Observation of High Intensity Radiation by Satellites 1958 Alpha and Gamm *J. Jet Prop.* 28 pp 588-592
[28] Haffner J W 1967 *Radiation and shielding in space* (New York and London: Academic Press) chapter IV p 112
[29] Rocco E M 2015 Gravitational disturbances generated by the Sun, Phobos and Deimos in orbital maneuvers around Mars with automatic correction of the semi-major axis *J. Phys.: Conf. Series* **641** 012027

[30] Baig S, McInnes C R 2010 Light-Levitated Geostationary Cylindrical Orbits Are Feasible *J. Guid. Control Dyn.* **33** pp 782-93
[31] Fang J *at al* 2011 Orbits of near-earth asteroid triples 2001SN263 and 1994 cc: Properties, origin, and evolution *Astron. J.* **141** pp 154-168
[32] Prado A F B A 2014 Mapping orbits around the asteroid 2001SN263 *Adv. Space Res.* **53** pp 877–889
[33] Lucking C *et al* 2011 A passive de-orbiting strategy for high altitude CubeSat missions using a deployable reflective balloon *8th IAA Symp. on Small Satellites* (Germany)
[34] Guerman A, Smirnov G 2012 Orbital maneuvres with single-input control *Advances in the Astronautical Sciences* **145** pp 171–181
[35] Deienno R *et al* 2016 Satellite de-orbiting via controlled solar radiation pressure *Cel. Mech. Dyn. Ast.* **126**(4) pp 433–59
[36] Rocco E M 2008 Perturbed orbital motion with a PID control system for the trajectory (Águas de Lindoia: XIV Colóquio Brasileiro de Dinâmica Orbital) pp 17-21
[37] Rocco E M 2012 Controle de trajetória com propulsão contínua para missões do tipo drag-free (Maranhão: Congresso Nacional de Engenharia Mecânica)
[38] Ma X, Li J 2013 Distant quasi-periodic orbits around Mercury *Astrophys. space Sci* **343 (1)** 83
[39] Tresaco E, Eipe A, Carvalho J P S 2016 Frozen Orbits for a Solar Sail Around Mercury *J. Guid. Control. Dynam.* **39**(7) pp 1659-1666
[40] Mazarico E *et al* 2014 The gravity field, orientation, and ephemeris of Mercury from MESSENGER observations after three years in orbit *J. Geophys. Res. Planets* **119** pp 2417-2436
[41] Mooij I E 1994 The motion of a vehicle in a planetary atmosphere *Faculty of Aerospace Engineering* (Delft – Netherlands) pp 29-31
[42] Benavoli A, Chisci L, Farina A 2007 Tracking of a Ballistic Missile with a-priori information *IEEE Transactions on Aerospace and Eletronic Systems* **43**(3) pp 1000-1016
[43] Fehlberg, E. 1968 Classical Fifth-, Sixth-, Seventh-, and Eighth-Order Runge Kutta Formulas with Stepsize Control *NASA Technical Report-287* (George C. Marshall Space Flight Center, Huntsville, Alabama – USA)
[44] Rocco E M, 2015 Gravitational disturbances generated by the Sun, Phobos and Deimos in orbital maneuvers around Mars with automatic correction of the semi-major *J. Phys.: Conf. Series* **641**
[45] Bate R R, Mueller D D, White J E 1971 *Fundamentals of Astrodynamics* Dove Public. Inc.
[46] Cunha A G, Takahashi R, Antunes C H 2012 *Manual de Computação Evolutiva e Metaheurística* (Editora Universidade Federal de Minas Gerais)
[47] Rocco E M; Souza M L O; Prado A F B A 2005 Multiobjective optimization applied to satellite constellations II: initial applications of the smallest loss criterion. *In 4th international workshop on satellite constellations and formation flying* (National Institute for Space Research, São José dos Campos, Brazil) pp 14–16
[48] Rossi A, Valsecchi G B 2006 Collision risk against space debris in Earth orbits *Cel. Mech. Dyn. Astron.* (**95**) pp 345–356
49] Smirnov N N 2002 *Space Debris Hazard Evaluation and Mitigation* (Taylor & Francis, London New York)
[50] Jesus A D C *et al* 2012 Evasive Maneuvers in Space Debris Environment and Technological Parameters *Mathematical Problems in Engineering* p. 15
[51] Carvalho J P S, Vilhena de Moraes R, Prado A F B A 2010 Some orbital characteristics of lunar artificial satellites *Cel. Mech. Dyn. Ast.* **108** pp 371-388
[52] De Saedeleer B, Henrard J 2005 The combined effect of J2 and C22 on the critical inclination of a lunar orbiter *Adv Space Res* **37** pp 80-87
[53] Chen J Y, Ning J S, Zhang C Y, Luo J 2005 On the determination of the lunar gravity field from the first Chinese lunar prospector mission *Chinese J. Geophys.* **48** pp 303-11

[54] Schubert G *et al* 2004 Interior Composition, Structure and Dynamics of the Galilean Satellites Jupiter: The Planet, Satellites and Magnetosphere *vol 1 ed Bagenal F* (Cambridge: University Press) chapter 13 pp 281-306
[55] Chapman C R 1996 S-Type Asteroids, Ordinary Chondrites, and Space Weathering: The Evidence from Galileo's Fly-bys of Gaspra and Ida *Meteoritics & Planetary Science* **31**(6) pp 699–725
[56] Hodges D H, Pierce G A 2002 *Introduction to structural dynamics and aeroelasticity* (New York – Cambridge University Press)
[57] Mukhopadhyay V 2003 Historical perspective on analysis and control of aeroelastic responses *Journal of Guidance, Control and Dynamics* **26**(5) pp 673–684
[58] Brejão L F, Brasil R M L R F 2016 Modeling a Rocket Elastic Structure as a Beck's Column Under Follower Force *In: Proceedings of CILAMCE 2016 – Brasília, Brazil*
[59] Goebel D M, Katz I 2008 Fundamentals of Electric Propulsion: Ion and Hall Thrusters (*JPL Space Sci. Tech. Series*)
[60] Raitses Y, Fisch N J 2001 Parametric investigations of a nonconventional Hall thruster *Phys. Plasmas* **8**(5) 2579
[61] Garrigues L *et al* 2008 Simulations of a Miniaturized Cylindrical Hall Thruster *IEEE Trans. Plasma Sci.* **36** 2034
[62] Raitses Y, Merino E, Fisch N J 2010 Cylindrical Hall thrusters with permanent magnets *J. Appl. Phys.* **108** 093307
[63] Scheeres D J, Williams B G, Miller J F 2000 Evaluation of the dynamic environment of an asteroid: Applications to 433 Eros *Journal of Guidance, Control and Dynamics* **23** (3) pp 466-475
[64] Zeng X Y, Gong S, Li J, Alfriend K 2016 Equilibrium points of elongated celestial bodies as the perturbed rotating mass dipole J*ournal of Guidance, Control and Dynamics* **39**(6) pp 1223-2016
[65] Tracy M B *et al* 2015 Physical modeling of triple near-earth asteroid (153591) 2001SN263 from radar and optical light curve observations *Icarus* **248** 499515
[66] Santos L B T, Prado A F B A, Sanchez D M 2017 Equilibrium points in the restricted synchronous three-body problem using a mass dipole model *Astrophysics and Space Science* **362**(3) p 61
[67] Sukhanov A A *et al* 2010 The Aster project: Flight to a near-Earth asteroid *Cosmic research* **48**(5), pp 443-455
[68] Ferreira J L *et al* 2009 Permanent Magnet Hall Thruster for Satellite Orbit Maneuvering With Low Power *Advances in Space Research* Vol. 1 pp 166-182
[69] Jesus A D C *et al* 2012 Evasive Maneuvers in Space Debris Environment and Technological Parameters *Mathematical Problems in Engineering* **2012** p 15
[70] Jesus A D C, Sousa R R, Neto E V 2015 Evasive Maneuvers in Route Collision with Space Debris Cloud *J. Phys.: Conf. Series* **641** 012021
[71] Sampaio J C *et al* 2014 Resonant Orbital Dynamics in LEO Region: Space Debris in Focus *Mathematical Problems in Engineering* **2014**
[72] Stern S A 1684 A rectilinear guidance strategy for short orbital transfers *Journal Spacecraft and Rockets* **21**(6) pp 542–545
[73] Humi M 1993 Fuel-optimal rendezvous in a general central force fiel**d** *Journal of Guidance, Control, and Dynamics* **16**(1) pp 215–217
[74] Vulpetti G, Johnson L and Matloff G L 2015 *Solar Sails: a novel approach to interplanetary travel* 2nd edition (New York: Copernicus Books) p 277
[75] Meireles L G 2016 *Análise de Transferências Orbitais com uso de Velas Solares* (Belo Horizonte, MG – Brasil)
[76] Mcinnes C R 2004 *Solar Sailing: Technology, Dynamics and Mission Applications* 2nd edition (London: Springer Praxis Books) p 296
[77] Marcy G W 2014 ApJ S 21020 (Preprint 1401.4195)
[78] Petigura E A 2016 ApJ 81836 (Preprint 1511.04497)

IOP Conf. Series: Journal of Physics: Conf. Series **911** (2017) 011002 doi:10.1088/1742-6596/911/1/011002

Peer review statement

All papers published in this volume of *Journal of Physics: Conference Series* have been peer reviewed through processes administered by the proceedings Editors. Reviews were conducted by expert referees to the professional and scientific standards expected of a proceedings journal published by IOP Publishing.

IOP Conf. Series: Journal of Physics: Conf. Series **911** (2017) 012001 doi:10.1088/1742-6596/911/1/012001

Searching for orbits around the triple system 45 Eugenia

B Y P M Mescolotti[1], A F B A Prado[1], A P M Chiaradia[2]andV M Gomes[2]

[1] National Institute for Space Research (INPE), São José dos Campos, SP, Brazil
[2] São Paulo State University (UNESP), School of Engeneering, Guaratinguetá, SP, Brazil

brunamasago@gmail.com

Abstract. Asteroids are small bodies that raises high interest, because they have unknown characteristics. The present research aims to study orbits for a spacecraft around the triple asteroid 45 Eugenia. The quality of the observations made by the spacecraft depends on the distance the spacecraft remains from the bodies of the system. It is used a semi-analytical model that is simple but able to represent the main characteristics of that system. This model is called "Precessing Inclined Bi-Elliptical Problem" (PIBEP). A reference system centered on the main body (Eugenia) and with the reference plane assumed to be in the orbital plane of the second more massive body, here called Petit-Prince, is used. The secondary bodies are assumed to be in elliptical orbits. In addition, it is assumed that the orbits of the smaller bodies are precessing due to the presence of the flattening of the main body (J_2). This work analyzes orbits for the spacecraft with passages near Petit-Prince and Princesses, which are the two smaller bodies of the triple system.

1. Introduction

According to [1], asteroids are bodies that orbit the Sun, but are too small to be considered planets. They may receive different classifications according to their orbital, physical, chemical and mineralogical characteristics. Most of the asteroids in the Solar System are located between the orbits of the planets Mars and Jupiter. This region is designated as the main asteroid belt.

Several missions have had or have an asteroid as the main target. Near-Shoemaker is designed to study the 433 Eros asteroid. In 1996, it collected images from 253 Mathilde. In 2000, after the approach, it landed on the asteroid 433 Eros [2]. In 2005, Hayabusa collected images and landed on the asteroid 24143 Itokawa [3]. Dawn visited, between 2011 and 2012, the asteroid 4 Vesta. In 2015, it reached 1 Ceres [4]. The Osiris-Rex spacecraft was launched in September 2016 and is expected to reach the asteroid Bennu (formerly called the 1999 RQ36) in 2018. The objective of the Osiris-Rex spacecraft is to collect materials from this asteroid and return them to Earth in 2023 [5].

2. The triple system 45 Eugenia

The asteroid 45 Eugenia was discovered at the Paris Observatory on June 27, 1857, by the astronomer Hermann Mayer Salomon Goldschmidt. The system consists of a main body (Eugenia) with 217 km in diameter and two smaller bodies (Petit-Prince and Princesse) with diameters of 5 and 7 km, respectively [6]. In Table 1, the characteristics of the triple asteroid are shown.

IOP Conf. Series: Journal of Physics: Conf. Series **911** (2017) 012001 doi:10.1088/1742-6596/911/1/012001

Table 1. Physical and orbital characteristics of the Eugenia system [6].

Asteroid	Main body	*a*	*e*	*i*	Period	Radius (km)	Mass (kg)
Eugenia	Sun	2.721 AU	0.083	6.61°	4.49 years	108.5	$5{,}63 \times 10^{18}$
Petit-Prince	Eugenia	1164.5 km	0.006	9°	4.7 days	3.5	$2{,}50 \times 10^{14}$
Princesse	Eugenia	610 km	0.069	18°	1.8 days	2.5	$2{,}50 \times 10^{14}$

3. Description

The objective of this work is to search for orbits around the main body of the system, with the goal that they can be used to observe all the bodies of the system. The first step is to choose the initial conditions for the orbits of the spacecraft that allow close passages by the asteroids. It is used the same technique developed in [7, 8]. After finding the initial conditions, the orbits are numerically integrated using the mathematical model called PIBEP [7, 8]. This model uses a reference system with origin on the main body (Eugenia) and have a reference plane in the orbital plane of Petit-Prince. The secondary bodies are assumed to be in elliptical orbits precessing due to the flattening of the main body (J_2). The spacecraft moves under the gravity forces of the three bodies involved in the system and the J_2 term of the gravity field of the main body. The evolution of the distances between the spacecraft and all the bodies of the system is monitored, and the time the vehicle remains close to those bodies is computed and shown in the results.

In a first analysis, a study of the eccentricities of the orbits is performed for all orbits obtained. Those with eccentricity greater than 1 are discarded, since they are hyperbolic orbits, therefore, not allowing repeated passages between the spacecraft and the secondary bodies. A second analysis is made to verify the distances between the spacecraft and the Eugenia main body (R1), between the spacecraft and Petit-Prince (R2), and between the spacecraft and Princesse (R3), all of them as a function of time. After choosing initial orbits around the central body, it is also eliminated orbits ending in collisions with one of the bodies. After that, the remaining trajectories are numerically integrated using the model PIBEP. This process also takes into account other aspects of the four-body dynamics, such as close approaches [9, 10] and gravitational captures [11, 12].

4. Mathematical model

The model used in the present research is the one shown in [6, 7]. It was noted that, in this system, the effects of the solar radiation pressure was not so important, as in the $2001SN_{263}$ [13] studied in [7], due to the larger masses of the bodies, which generates higher gravitational forces. The distances between the spacecraft and Eugenia (R1), spacecraft and Petit-Prince (R2) and spacecraft and Princesse (R3) are given by Eqs. 1-3, where (x_1, y_1, z_1) are the coordinates of Eugenia, (x_2, y_2, z_2) are the coordinates of Petit-Prince and (x_3, y_3, z_3) are the coordinates of Princesse.

$$R_1 = \sqrt{(x - x_1)^2 + (y - y_1)^2 + (z - z_1)^2} \tag{1}$$

$$R_2 = \sqrt{(x - x_2)^2 + (y - y_2)^2 + (z - z_2)^2} \tag{2}$$

$$R_3 = \sqrt{(x - x_3)^2 + (y - y_3)^2 + (z - z_3)^2} \tag{3}$$

The equations of motion for the spacecraft in the inertial system, considering the above assumptions, are given by Eqs. 4-6, shown below:

$$\ddot{x} = -\mu_2 \frac{(x - x_2)}{r_1{}^3} - \mu_3 \frac{(x - x_\gamma)}{r_2{}^3} - \mu_1 \frac{(x - x_\alpha)}{r_3{}^3} - \mu_1 J_2 r_1{}^2 \left(\frac{3x}{2r^5} - \frac{15z^2x}{2r^7}\right) \tag{4}$$

$$\ddot{y} = -\mu_2 \frac{(y - y_2)}{r_1{}^3} - \mu_3 \frac{(y - y_3)}{r_2{}^3} - \mu_1 \frac{(y - y_1)}{r_3{}^3} - \mu_1 J_2 r_1{}^2 \left(\frac{3y}{2r^5} - \frac{15z^2y}{2r^7}\right) \tag{5}$$

$$\ddot{z} = -\mu_2 \frac{(z - z_2)}{r_1{}^3} - \mu_3 \frac{(z - z_3)}{r_2{}^3} - \mu_1 \frac{(z - z_1)}{r_3{}^3} - \mu_1 J_2 r_1{}^2 \left(\frac{9z}{2r^5} - \frac{15z^3}{2r^7}\right) \tag{6}$$

where $J_2 = 0{,}06$ [6] is the flatness of the central body and r = 108.5 km is the value of the radius of Eugenia. The gravitational parameters of Eugenia, Petit-Prince and Princesse are given by μ_1, μ_2 and μ_3, respectively.

5. Results

The evolution of the distances between the spacecraft and the bodies of the system was analyzed in two intervals: up to 50 km and 50-100 km. The simulations considered only the situation where the asteroids start their motion at the periapsis of their orbits. The next figures illustrate the evolution of the distances between the spacecraft and all the bodies of the system. The term resonance is used in the sense of the instantaneous initial orbit only. It means that the trajectory obtained from a given resonance would have several encounters with the resonant body, if the dynamics was the "two-body problem". But, since the dynamics is much more complex, the trajectory is modified and real resonances may not occur. In that sense, it is just a technique used to find initial conditions, which proofed to work very well. See [7, 8] for more explanations. A large number of resonances was simulated, and the best trajectories are shown next.

5.1. *Internal orbit to Petit-Prince, initially in resonance 1:2, with the spacecraft starting at the apoapsis of its orbit*

In this orbit, the spacecraft does not observe the main body Eugenia from the distances limit defined (up to 50 km and 50-100 km). In the interval between 50-100 km it remains 4.02 days observing Petit-Prince. The time is not long, but there are many passages, which can be useful for the observations. Regarding Princesse, it is noted that the spacecraft remains less than 1 day close to the body, but also making several passages. Figure 1 shows the evolution of the distances, as a function of time, between the spacecraft and the bodies. Figures 2 and 3 show a zoom of the passages of the spacecraft in the intervals simulated.

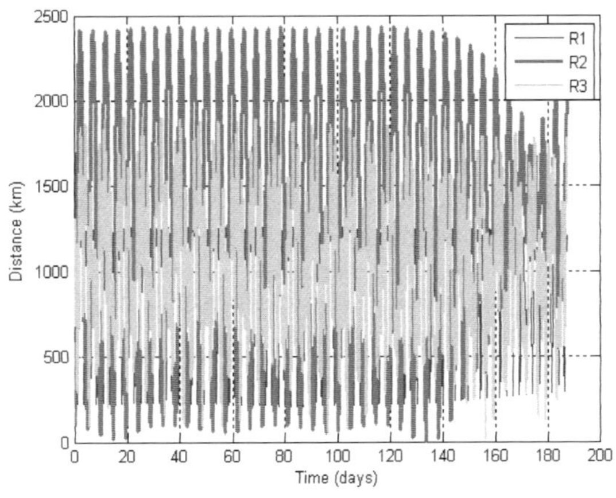

Fig. 1. Distances of the spacecraft as a function of time until Eugenia (blue), Petit-Prince (red) and Princesse (green).

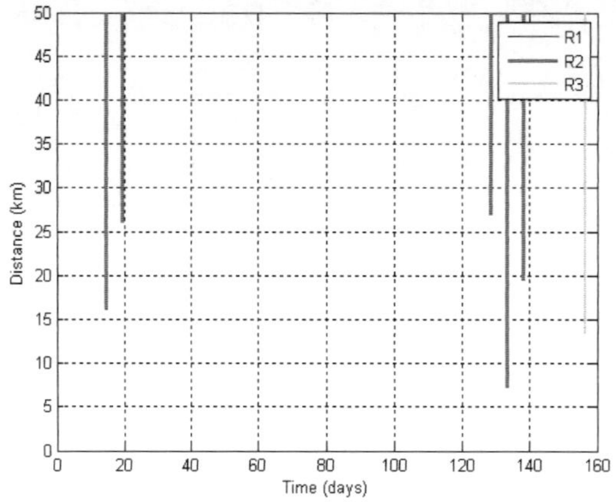

Fig. 2. Distances of the spacecraft as a function of time until Eugenia (blue), Petit-Prince (red) and Princesse (green) in the interval less than 50 km.

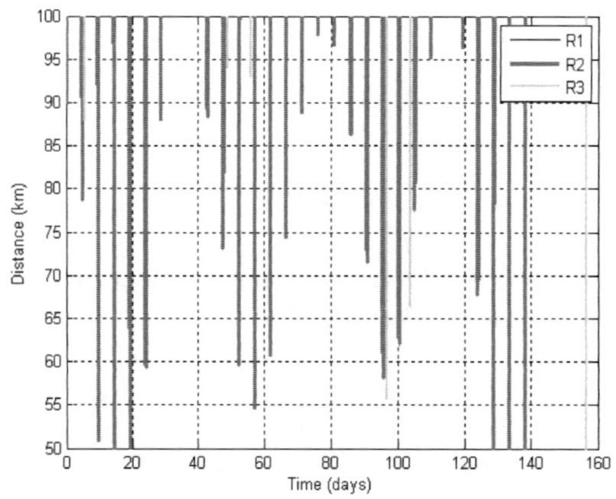

Fig. 3. Distances of the spacecraft as a function of time until Eugenia (blue), Petit-Prince (red) and Princesse (green) in the interval 50-100 km.

5.2. *Internal orbit to Princesse initially in resonance 5:9, when the spacecraft starts its motion at the periapsis of its orbit*

The spacecraft does not observe the main body Eugenia and Petit-Prince. When the spacecraft is in its periapsis, Princesse is observed for 0.40 days at a distance of up to 50 km and in the range 50-100 km for 3.64 days, with several passages. Figure 4 shows the evolution of the distances, as a function of time, between the spacecraft and the three bodies. Figures 5 and 6 show a zoom of the passages of the spacecraft in the simulated intervals.

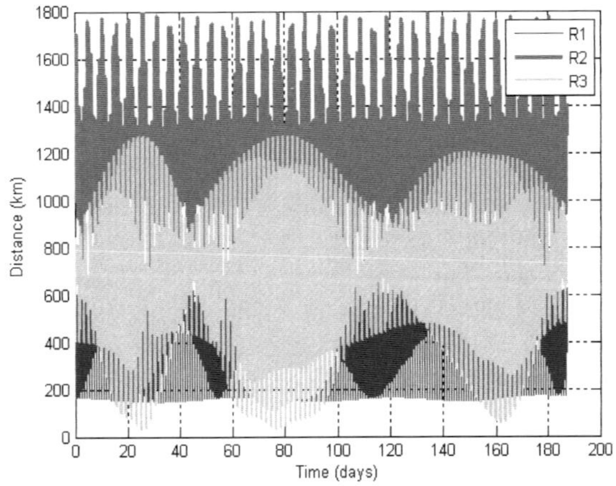

Fig. 4. Distances of the spacecraft according to the time until Eugenia (blue), Petit-Prince (red) and Princesse (green).

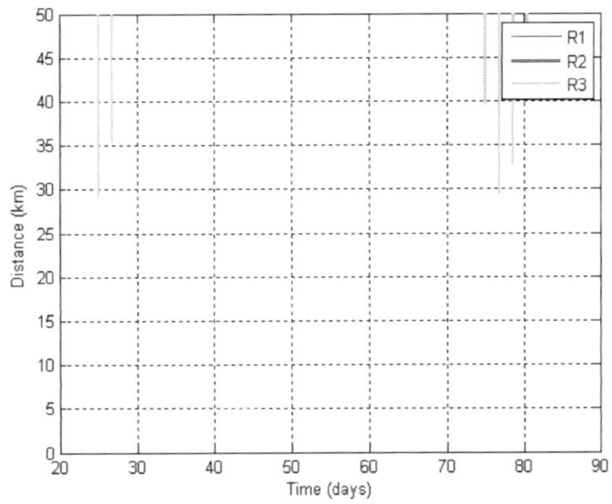

Fig. 5. Distances of the spacecraft according to the time until Eugenia (blue), Petit-Prince (red) and Princesse (green)in the interval less than 50 km.

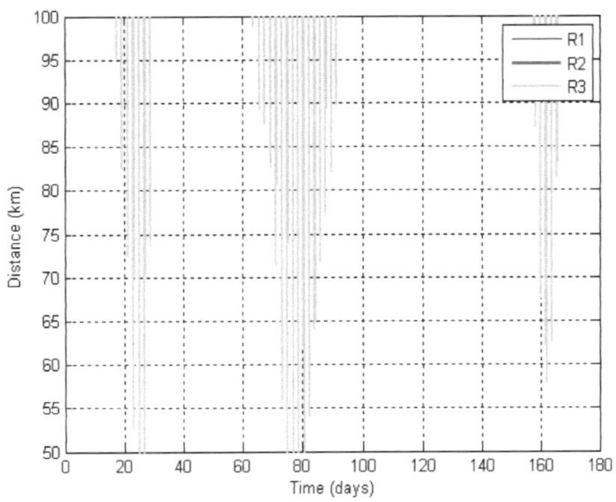

Fig. 6. Distances of the spacecraft according to the time until Eugenia (blue), Petit-Prince (red) and Princesse (green)in the interval 50-100 km.

6. Conclusions

Based in the results shown in the present paper, it is noted that, using initially internal orbits in resonance with Petit Prince, the spacecraft had no close encounter with the central body, while using orbits which are initially in internal resonance with Princesse, the spacecraft did not observe the main body and Petit-Prince.

It means that the strategy to search for orbits developed in [7] does not give orbits with large observational times, as found for the triple system $2001SN_{263}$. The reason is that the smaller bodies of the 45 Eugenia system are very far from the central body, generating close encounters between the spacecraft and the bodies of the system, but with shorter durations. Those passages are important, even if the time is not too long. It also means that it is necessary to use more orbital maneuvers to paste together the different orbits found with the present technique, since no adequate orbits were found to observe all the bodies in a single trajectory. On the otherside, the distances between the bodies are larger, and this implies in a system much less disturbed, so the orbits are more stable in general, requiring less fuel consumption on station-keeping maneuvers required to compensate effects of the perturbations.

Acknowledgments

The authors wish to express their appreciation for the support provided by grants #304700/2009-6, 472352/2013-0 and 473164/2013-2 from the National Council for Scientific and Technological Development (CNPq); grants #2011/08171-3, 2011/13101-4 and 2016/14665-2, from São Paulo Research Foundation (FAPESP), and the financial support from the National Council for the Improvement of Higher Education (CAPES).

References

[1] Araújo R A N 2011 O sistema triplo de asteroides 2001SN263: dinâmica orbital e estabilidade. Doctoral Thesis in Space Engineering and Technology INPE
[2] NASA Science 2016 Available in <http://science1.nasa.gov/missions/near>. Acess in: 06 nov.
[3] Yoshikawa M, Fujiwara A, Kawaguchi J 2007 Hayabusa and Its Adventure Around the Tiny Asteroid Itokawa *Highlights Astron.* **14** 323
[4] Russell C T, Raymond C A 2011 The Dawn Mission to Vesta and Ceres *Space Sci. Rev.* **163** 3
[5] NASA 2016 Available in <https://www.nasa.gov/osiris-rex>. Acess in: 12 nov.
[6] Marchis F, Lainey V, Descamps P, Berthier J, Van Dam M, de Pater I, Macomber B, Baek M, Le Mignant D, Hammel H B, Showalter M, Vachier F 2010 A dynamical solution of the triple asteroid system (45) Eugenia *Icarus* **210** 635
[7] Masago BYP L2014 Estudo de órbitas ressonantes no sistema triplo $2001SN_{263}$ Master's Thesis in Space Engineering and Technology INPE
[8] Masago B Y P L, Prado A F B A, Chiaradia A P M, Gomes V M 2016 Developing the "Precessing Inclined Bi-Elliptical Four-Body Problem with Radiation Pressure" to search for orbits in the triple asteroid $2001SN_{263}$ *Adv. Space Res.* **57**962
[9] Broucke R A, Prado A F B A 1993 Jupiter Swing-By trajectories passing near the Earth *Spacef. Mec.* **82** 1159
[10] Gomes V M, Prado A F B A, Golebiewska J 2013 Dynamics of space particles and spacecrafts passing by the atmosphere of the Earth *Scient. World J.*
[11] Neto E V, Prado A F B A 1998 Time-of-flight analyses for the gravitational capture maneuver *J. Guid. Cont. Dyn.* **21** 122
[12] Prado A F B A 2002 Numerical study and analytic estimation of forces acting in ballistic gravitational capture *J. Guid. Cont. Dyn.* **25** 368
[13] Araújo R A N, Winter O C, Prado A F B A, Sukhanov A 2012 Stability regions around the components of the triple system $2001SN_{263}$ *Mont. Not. Royal Astr. Soc.* **423** 3058

IOP Conf. Series: Journal of Physics: Conf. Series **911** (2017) 012002 doi:10.1088/1742-6596/911/1/012002

Fractal boundaries in chaotic hamiltonian systems

R L Viana[1], A C Mathias[1], F A Marcus[1], T Kroetz[2] and I L Caldas[3]

[1] Departamento de Física, Universidade Federal do Paraná, Curitiba, Paraná, Brazil
[2] Departamento de Física, Universidade Tecnológica Federal do Paraná, Pato Branco, Paraná, Brazil
[3] Instituto de Física, Universidade de São Paulo, São Paulo, São Paulo, Brazil

E-mail: viana@fisica.ufpr.br

Abstract. Fractal structures are typically present in the dynamics of chaotic orbits in non-integrable open Hamiltonian systems and result from the extremely complicated nature of the invariant manifolds of unstable periodic orbits. Exit basins, the set of initial conditions leading to orbits escaping through a given exit, have very frequently fractal boundaries. In this work we analyze exit basin boundaries in a dynamical system of physical interest, namely the motion of charged particles in a magnetized plasma subjected to electrostatic drift waves, and characterize in a quantitative way the fractality of these structures and their observable consequences, as the final-state uncertainty.

1. Introduction

In open chaotic hamiltonian systems there are initial conditions generating orbits which eventually exit from the system. Suppose there are two or more different ways in which an orbit can exit from the system. We define the exit (or escape) basin as the set of initial conditions in a certain phase-space region leading to orbits escaping through a given exit [1]. In particular we are interested in characterizing the boundary between these exit basins, which is often fractal and thus presents final-state sensitivity: a slight perturbation of some initial condition may lead to an orbit escaping through a different exit, impairing our capacity of predict the future outcome of the system [2, 3].

This problem has various applications in many fields. In celestial mechanics for example, Contopoulos and coworkers [4, 5] has studied the escape of stars from a galaxy using a paradigm hamiltonian such that, for energies larger than a critical value, there exists channels whereby orbits may escape to infinity. More recently Aguirre *et al.* have studied the exit basins of such a problem using the Hénon-Heiles hamiltonian [6].

Other field in which exit basins have been intensively studied in recent years is plasma physics. The escape of magnetic field lines in a magnetically confined fusion plasma with destroyed magnetic surfaces presents fractal exit basins [7, 8]. Another problem in which such fractal structures appear is the charged particle motion in a magnetized plasma under electrostatic drift waves [9], which is a relevant scenario in the study of plasma turbulence in the plasma edge in a tokamak [10, 11].

In this paper we present some results on the characterization of exit basins in the abovementioned problem of particle motion under two drift waves, which is hamiltonian but generally nonintegrable. The existence of a chaotic, area-filling orbit in phase space, and its corresponding homoclinic tangle, is ultimately responsible for the fractality of exit basins [12]. We characterize the latter by computing the box-counting dimension of the exit basin boundaries, using the uncertainty fraction method [2] and also by calculating the basin entropy and basin boundary entropy, two concepts recently developed by Daza et al. [13].

This paper is organized as follows: in Section II we introduce the physical system to be considered, as well as the chaoticity of particle motion we focus on. Section III deals with the subject of exit basins and the characterization of their fractality. The last Section is devoted to our Conclusions.

2. Model

It is well-known that the motion of a particle of mass m and charge e in a uniform magnetic field B_0, also called gyration, is a combination of a circular motion of radius mv/eB_0 on the plane perpendicular to $\mathbf{B}$ and a uniform motion along $\mathbf{B}$. The motion of particle's guiding center suffers a drift when an electric field $\mathbf{E} = -\nabla\Phi$ is present. The velocity of such $\mathbf{E}\times\mathbf{B}$ is given by [14]

$$\mathbf{v} = -\frac{\nabla\Phi \times \mathbf{B}}{B^2}. \tag{1}$$

In our discussion we neglect all energy dissipation caused by, e.g. collisions between particles with energy and momentum exchange. If such effects were to be taken into account the system would become dissipative, with a completely different dynamics. However, the effect of such collisions can be neglected if the collisional frequency is small enough.

Factoring out the gyration, we can focus on the drift motion of the charged particles with a velocity $\mathbf{v}$, whose components satisfy canonical equations

$$\frac{dx}{dt} = -\frac{\partial H}{\partial y}, \qquad \frac{dy}{dt} = \frac{\partial H}{\partial x}, \tag{2}$$

where the Hamiltonian is $H(x, y, t) = \Phi(x, y, t)/B_0$. In the following, all variables have been suitably nondimensionalized.

The electric potential has an equilibrium part $\Phi_0(x)$ corresponding to a radial electric field and a perturbation caused by two electrostatic drift waves with amplitudes $A_{1,2}$, frequencies $\omega_{1,2}$ and wave vectors $\mathbf{k}_{1,2}$, in which, by a canonical transformation, leads to the following Hamiltonian [15]

$$H(x, y, t) = \Phi_0(x) - u_1 x + A_1 \sin(k_{x1}x)\cos(k_{y1}y) + A_2 \sin(k_{x2}x)\cos[k_{y2}(y - ut)], \tag{3}$$

where $u_{1,2} = \omega_{1,2}/k_{y1,2}$ are the phase velocities of the drift waves and $u = u_2 - u_1$. We suppose that a stationary wave pattern along the radial distance x and a travelling wave along the poloidal direction y, such that the wave numbers are chosen as $k_x = n\pi/L_x$ and $k_y = 2\pi m/L_y$, where m and n are suitably chosen positive integers and $L_{x,y}$ are characteristic lengths [9]. We suppose a uniform electric field applied in the radial direction of the tokamak, such that $\Phi_0 = E_0 x$. In this model we neglect some collective effects in the plasma and consider it simply a collection of non-interacting charged particles. Within this approximation, the interaction of the confined plasma with the electric and magnetic fields does not lead to collective effects, such as mechanical or magnetohydrodynamical waves.

The canonical equations (2) were numerically integrated for the Hamiltonian (3), and a Poincaré map was taken after considering only those values of (x, y) at times equal to an integer multiple of the period $T = 20\pi$. We choose A_2 (amplitude of the second wave) as the perturbation strength. The case $A_2 = 0$ represents an integrable system, and the particle motion is bounded within 4×4 cells limited by the separatrices emanating from the hyperbolic

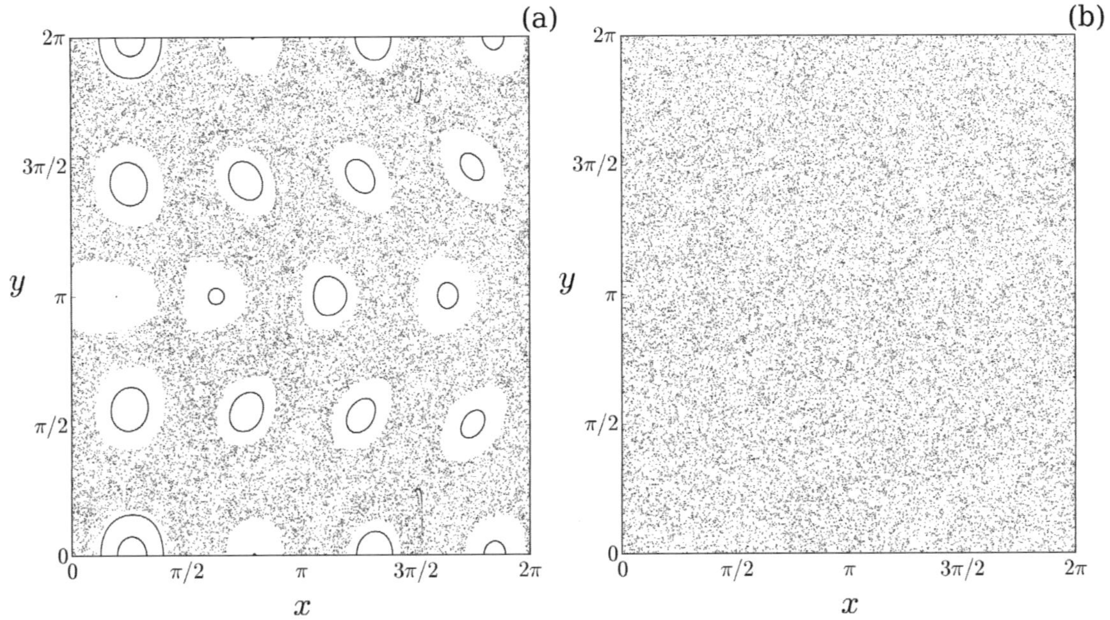

Figure 1. Phase portrait for the Poincaré map for $\omega_1 = \omega_2 = 0.2$, $k_{x1} = k_{y1} = 2.0$, $k_{x2} = \sqrt{3}$, $k_{y2} = 1.0$, $u = 0.1$, $A_1 = 1.0$ and (a) $A_2 = 0.8$, (b) $A_2 = 1.5$.

fixed points of the Poincaré map. In the interior of each cell there are closed curves orbiting around the elliptic fixed points.

In Fig. 1 we show two phase portraits for the Poincaré map obtained with different values of the perturbation intensity. For the case of smaller perturbation [Fig. 1(a)] the system integrability is broken, and the separatrices of the cells give place to a localized area-filling chaotic orbit, which is considerably enlarged as the perturbation is increased, practically engulfing all the remnant islands [Fig. 1(b)].

3. Exit basins

The chaotic orbits depicted in Fig. 1(a) and (b) allow particle transport over all the available phase space region. Since the left-hand boundary of this region coincides with the tokamak wall ($x = 0$) those particles hitting this line can be considered as effectively lost, due to many plasma-wall processes like sputtering. In order to investigate the distribution of particle escape through this wall we will divide it into two congruent segments: $\mathbf{A} = \{x = 0, 0 < y \leq \pi\}$ and $\mathbf{B} = \{x = 0, \pi < y \leq 2\pi\}$.

The next step is to cover the phase space region in Fig. 1 with a fine mesh of points and, considering each mesh point an initial condition (x_0, y_0) we iterated the Poincaré map and wait until the ensuing orbit escape through the wall at $x = 0$. If the trajectory has escaped through exit **A** (**B**) the corresponding initial condition is painted green (red). The sets of green and red pixels are thus numerical approximations of the basins of exits **A** and **B**, respectively. The white pixels stand for initial conditions that do not escape, since they are inside the periodic islands. The results are depicted in Fig. 2(a) and (b) for the same parameter values used in the phase portraits of Fig. 1(a) and (b).

A visual inspection of Figs. 2(a) and (b) shows that the size of the green basin is considerably

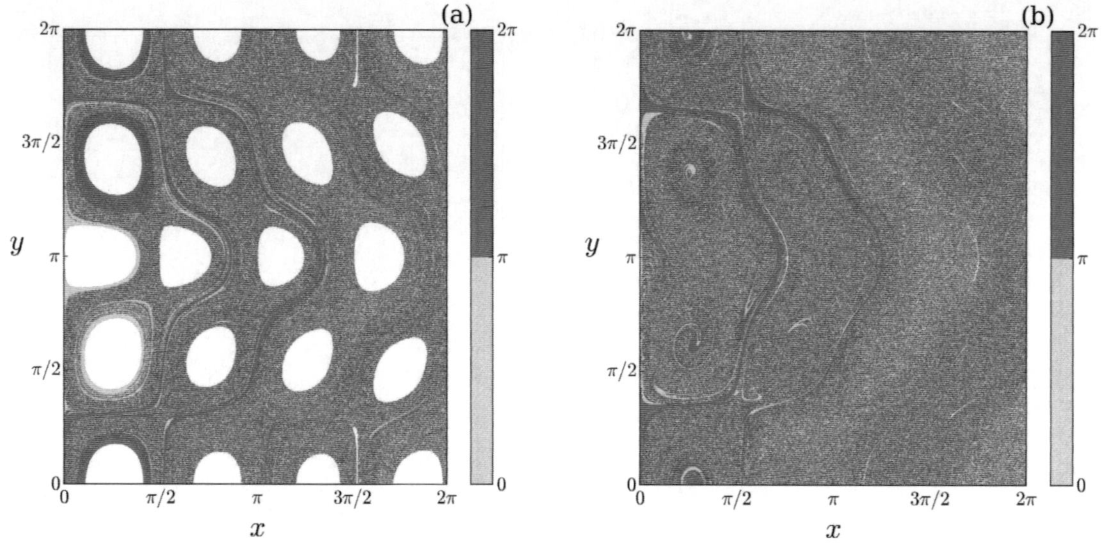

Figure 2. Basins of the exits $\mathbf{A} = \{x = 0, 0 < y \leq \pi\}$ (green pixels) and $\mathbf{B} = \{x = 0, \pi < y \leq 2\pi\}$ (red pixels) for the same parameters of Fig. 1, and (a) $A_2 = 0.8$, (b) $A_2 = 1.5$.

smaller than for the red basin. This means that most initial conditions escape through exit **B**, i.e. the escape pattern is strongly nonuniform. The reason for this behavior is the structure of invariant manifolds (stable and unstable) that cross themselves in an infinite number of homoclinic and heteroclinic points. Since the manifolds have an extremely involved behavior, the resulting meanders act as escape channels, through which the phase space trajectories are directed to escape towards $x = 0$.

When two exit basins are intertwined the way shown in Fig. 2, it is typically found that their common boundary is fractal. In order to check this fact we computed the box counting dimension of this boundary using the uncertainty fraction technique [2, 3]. We know that an initial condition is known up to a given uncertainty ε, represented by a ball of radius ε centered at that initial condition. If this ε-ball intercepts the basin boundary, we cannot be sure to which exit that initial condition will evolve to one. In this case we say that the initial condition is ε-uncertain.

The fraction of uncertain initial condition scales with the uncertainty as $f(\varepsilon) \sim \varepsilon^{\alpha}$, where $0 < \alpha < 1$ is called the uncertainty exponent. The latter is related to the box-counting boundary dimension D_0 by $\alpha = 2 - D_0$, in a two-dimensional phase space [2]. If the boundary is a smooth curve, then $\alpha = 1$ and $D_0 = 1$, whereas $\alpha < 1$ is for a fractal boundary. Using this technique we found that the box-counting dimension of the boundary between red and green basins in Fig. 2(a) and (b) is $D_0 = 1.945 \pm 0.002$ and $D_0 = 1.956 \pm 0.002$, respectively. Hence the basin boundary dimension has increased with the perturbation, and it approaches 2.0, which would mean an area-filling curve.

The box-counting dimension, while useful to characterize the fractality of the exit basin boundary, do not give information about the degree in which these basins are intertwined. A more convenient measure for this purpose is the basin entropy introduced by Daza *et al.* [13]. We start by covering a given region of the phase space by a grid of 1000×1000 points of each basin, distributed inside 4×10^4 boxes. For each grid point, a maximum of 1.5×10^4 stroboscopic map points were obtained, and excluding those initial conditions that have not escaped during

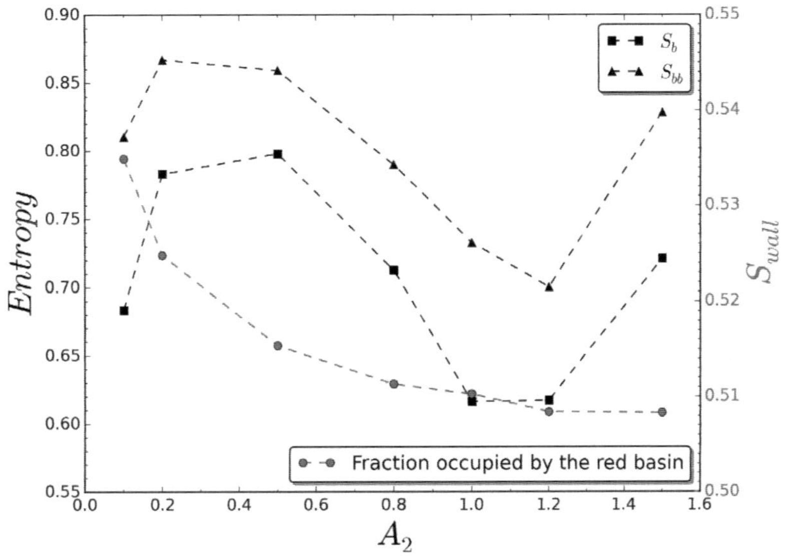

Figure 3. Basin entropy and basin boundary entropy for the case of two exits [see Fig. 2] as a function of the second wave amplitude. We also plot the fraction occupied by the red basin.

that time.

Let n_1 be the number of points of the box corresponding to the exit 1, and n_2 to exit 2. For each box we obtained the probabilities $p_1 = n_1/(n_1 + n_2)$ and $p_2 = n_2/(n_1 + n_2)$, and we apply the Gibbs entropy definition for each box [13]:

$$S = -\sum_{i=1}^{2} p_i \log_2(p_i). \tag{4}$$

The basin entropy S_b is the sum of the values of S for those boxes such that all the initial conditions escape, divided by the number of boxes N_b for which all initial conditions escape. The boundary basin entropy S_{bb} was obtained analogously, but excluding those boxes for which either $p_1 = 0$ or $p_2 = 0$, i.e., there were considered only the boxes that have at least one boundary in their interior.

If there were a single exit the basin entropy would vanish, meaning zero final-state uncertainty. On the other hand, if the two exits are equiprobable, in the sense that their basins are of comparable sizes and densely intertwined, the maximum value of the basin entropy would be $S_b = \log_2 2 = 1$, i.e. a completely randomized basin structure.

We show in Fig. 3 the values of the basin entropy S_b and the basin boundary entropy S_{bb} for the two exits as a function of the second wave amplitude A_2. It is also plotted the fraction S_{wall} occupied by the (red) basin of the exit **B** (the fraction of the green basin is $1 - S_{wall}$). The values of S_{wall} are slightly higher for weak perturbation (about 53%) and tend to 51% for strong perturbation. This fact is exemplified by Fig. 2(a) and (b), in which the green and red basins are more intertwined for stronger perturbation. We observe that the values of S_b fluctuate between 0.68 and 0.80 for the range of perturbation considered, meaning that the basins are mixed together in a very complicated way for all considered values of the perturbation strength A_2.

4. Conclusions

The dynamics in chaotic orbits of non-integrable open Hamiltonian systems is strongly influenced by fractal structures coming from the complicated intersections between stable and unstable invariant manifolds of unstable periodic orbits embedded in the chaotic orbit. In particular, the chaotic saddle (formed by the intersections between the manifolds) has a Cantor dust structure similar to the horseshoe set. Such fractal structures have observable consequences in open Hamiltonian systems of physical interest, like the motion of charged particles in a magnetic field under the influence of electrostatic drift waves of different frequencies and wave numbers. In this case, we analyzed the presence of fractal structures in the so-called exit basins, which are the sets of initial conditions leading to a escaping orbit through some specified exit. The boundary between two escaping basins has shown to be a fractal curve through direct computation of its box-counting dimension, showing a value close to the phase space dimension, which means a highly convoluted curve. This fact leads, as a consequence, to a final-state uncertainty, i.e. once the initial condition is determined within a given uncertainty, it turns to be increasingly difficult to know for sure to which exit the ensuing orbit asymptotes to. We verified this fact also by computing the basin entropy, which is a novel measure for the degree of uncertainty related to fractal exit basin boundaries. Our results are not restricted to this particular physical system though, and are quite often observed in other open Hamiltonian systems as well.

Acknowledgments

This work has received the partial financial support from the following Brazilian government agencies: CNPq, CAPES and FAPESP (Processo 2011/19296-1). We would like to acknowledge useful discussions with Prof. Miguel Sanjuán (Madrid).

References

[1] S. Bleher, C. Grebogi, E. Ott, and R. Brown, Phys. Rev. A **38** (1988) 930.
[2] S. W. McDonald, C. Grebogi, E. Ott, and J. A. Yorke, Physica D **17** (1985) 125.
[3] J. Aguirre, R. L. Viana, and M. A. F. Sanjuán, Rev. Mod. Phys. **81** (2009) 333.
[4] G. Contopoulos, Astron. Astrophysics **231** (1990) 41.
[5] G. Contopoulos, H. E. Kandrup and D. Kauffman, Physica D **64** (1993) 310.
[6] J. Aguirre, J. C. Vallejo, and M. A. F. Sanjuán, Phys. Rev. E **64** (2001) 066208.
[7] E. C. da Silva, I. L. Caldas, R. L. Viana, and M. A. F. Sanjuán, Phys. Plasmas **9** (2002) 4917.
[8] J. S. E. Portela, I. L. Caldas, R. L. Viana and M. A. F. Sanjuán, Int. J. Bifurcat. Chaos **17** (2007) 4067.
[9] W. Horton, Plas. Phys. Contr. Fusion **27** (1985) 937.
[10] W. Horton, *Turbulent transport in magnetized plasmas* (World Scientific, Singapore, 2012)
[11] A. C. Mathias, R. L. Viana, T. Kroetz, and I. L. Caldas, Physica A **469** (2017) 681.
[12] R. K. W. Roeder, B. I. Rapoport, and T. E. Evans, Phys. Plasmas **10** (2003) 3796.
[13] A. Daza, A. Wagemakers, B. Georgeot, D. Guéry-Odelin, and M. A. F. Sanjuán, Sci. Rep. **6** (2016) 3416.
[14] G. Schmidt, *Physics of high-temperature plasmas* (Academic Press, New York, 1979).
[15] F. A. Marcus, I. L. Caldas, Z. O. Guimarães-Filho, P. J. Morrison, W. Horton, Yu. K. Kuznetsov, and I. C. Nascimento, Phys. Plasmas **15** (2008) 112304

Deviations in CBERS-4 Satellite Direction Components From The Electromagnetic Disturbance of Communication Antennas

A Heilmann[1,2], **C Fernandes**[1]

[1] Center of Studies of the Sea, Federal University of Paraná, Curitiba, PR, Brazil.
[2] Laboratory of Geodesic, Federal University of Paraná, Curitiba, PR, Brazil.

[1]heilmann@ufpr.br

Abstract. The CBERS-4 is a low Earth orbit satellite, with a set of antennas S-band/UHF for communication almost omni-direcional. For the electromagnetic radiation from transmission antennas, was developed a model of electromagnetic disturbance considering the antennas theory and the laws of the conservation energy-momentum. Was propagated the orbit of the CBERS-4 satellite considering your state vector from the March 14, 2016, at 11^h 14^m 15.23^s using the equation of motion in the form of cartesian components. From the state vector of the CBERS-4 satellite was possible to propagate the orbit for different periods, without disturbance (considering just the problem of two bodies) and with a disturbance of electromagnetic origin. The model of reaction of electromagnetic acceleration on the satellite depends on only the type of antenna. Quadrifilar and parabolic propeller antennas were considered in this paper. Using the equation of motion of the satellite based on the method of Runge-Kutta of fourth and fifth degree, the effect disturber this modeling was applied on the CBERS-4 considering the mass of satellite, characteristics of antenna, power irradiated and gain maximum of antenna. The final analysis discusses the values of components in the direction (radial, cross and normal) and the coordinates X-Y-Z considering the case disturbed to both antennas.

1. Introduction

The disturbed orbital motion of artificial satellites is historically discussed, and still presents problems not solve about the nature of the disturbances, and some disturbances still not characterised or considered in their propagated orbit models. Among the disturbances known and cited in the literature, Cook (1962), Seeber (1993), Montenbruck and Gill (2005), Eshagh and Najafi (2007), Cojocaru (2007), cite the influence of the gravitational attraction of the Moon and the Sun, pressure of solar radiation direct and indirect (albedo), relativistic effects, disturbances with origin in the Earth's magnetic field, due to the wind solar, Poynting-Robertson effects and atmospheric drag. About the disturbance due to the electromagnetic effects from the antennas shipped in a GPS sattelite, Ziebart, Sibthorpe and Cross (2007) and Solano et. al. (2012), discuss about a constant of acceleration radial around 2.7×10^{-10} m/s^2, as one of the elements responsible for the disturbance on their respective orbitals elements.

Many researchers consider the electromagnetic disturbance associated with pressure of solar radiation, and apart from Maral and Bousquet (1998) – that ratify in a few lines, that the electromagnetic radiation of antennas creates a pressure that cannot be disregarded if the power of transmission is high – others authors that suggests the existence of the effects of linear momentum from the disturbance of origin electromagnetic, have not been found during the research of this work. According to Maral and Bousquet (1998), the force produced by the radiation of an antenna in a satellite with 1 kW of EIRP is 0.3 x 10^{-5} N, affirming that the disturbance is significant if the diagram of radiation of an antenna quadrifilar propeller, presents the gain and concentrated directivity. Zhou, Liu and He (2008) and also Bukhsh and Jonsson (2010), published articles to deal with analyze the standards of interference of the system of antennas in a satellite, having the purpose of explain the phenomena of electromagnetic interference on a set of antennas, not considering their effects about the presence of a linear momentum.

The CBERS-4 satellite is a satellite of low orbit with modules of camera for geometric and radiometric analysis. It has a total mass of 2080 kg, power generated of 2300 W and antennas type quadrifilar and parabolic propeller, operating in the frequency spectrum of the VHF and UHF and in the range of S-band. These antennas allow a communication almost omni-direcional, with the maximum of the radiation power spectrum oriented in the Z direction.

The objective of this research is to answer if the CBERS-4 low orbit satellite, with antennas transmitting electromagnetics signals (quadrifilar and parabolic propeller), admits a *momentum* when it sends an electromagnetic beam towards a receiving antenna in terrestrial ground. And in this case, observe the possible changes in the orbital elements, and from this to discuss the disturbances in the components: velocity, altitude, semi-major axis, orbit eccentricity, mean anomaly, longitude of ascending node, orbit slope, argument of perigee, X-Y-Z coordinates, and the radial, normal and tangential components of this satellite.

The normal component is perpendicular to the orbital plane and positive in the direction of the angular momentum vector. The radial orthogonal component is along the vector radius of the satellite and positive outside the central body, and the transverse component is in the direction of the orbital plane and positive in the direction of the movement of the satellite.

2. Electromagnetic disturbance model

The communication between satellite antennas and a receiving station on terrestrial soil suggests an analysis, considering the conservation of linear momentum, about the effects of electromagnetic waves at the moment of transmission, on the set of antenna plus satellite. In the presence of a continuous transmission of the satellite antenna, for the communication link to occur, the presence of a linear momentum on the satellite can result in a disturbance in its orbit.

The considerations of satellite mass, orbit trajectory and antenna parameters used in spatial communications can suggest a strong correlation with a mathematical model that describes the real perturbative effects in the orbit of satellites, which perform communication from antennas with gain and power characteristics (of the order of tens of Watts) of known transmission.

Using the Electromagnetic Reaction Acceleration Model (*Prop-Orbit Antenna*) developed by Heilmann et. al. (2012, 2013), that considers the electromagnetic momentum on a satellite, and calculates the acceleration of electromagnetic reaction on a satellite, and the principal parameters of antenna theory: antenna type and power of radiation, in a graphical platform generated in MatLab environment that allows to choose the satellite, considering its respective state vector and the period of propagation of the orbit in days (with steps of 1-30 minutes).

2.1 Parameters of the CBERS-4 satellite

The satellite orbit describes its position in the orbital plane at a given time t and it is calculated using the six orbital parameters, called orbital elements or Keplerian. These parameters describe the movement of a satellite around the Earth. The parameters are: semi-axis major (a) [m], eccentricity (e), slope of the orbital plane (i) [°], straight ascension of the ascending node (Ω) [°], argument of the perigee (ω)[°] e mean anomaly (M). Where (a) and (e) describes the shape and size of the orbit described by the satellite; (Ω), (i) and (ω) are the Euler angles that define the location of the orbital plane in space.

With the orbital elements of the CBERS-4 satellite is possible to obtain its corresponding parameters of the state vector, composed of Cartesian coordinates of position and velocity, in relation to the inertial reference system (Montenbruck and Gill, 2005) (table 1).

IOP Conf. Series: Journal of Physics: Conf. Series **911** (2017) 012003 doi:10.1088/1742-6596/911/1/012003

Table 1. Parameters of the CBERS-4 satellite.

1 40336U 14079A 16073.88076422 .00000047 00000-0 31380-4 0 9991[a]
2 40336 98.4979 150.5454 0001341 63.9076 296.2238 14.35445877 66370[a]

Mass (kg)	EIRP[b] (W)	Power[c] (W)	Inertial System (24 hours after)	
			Position (m)	Velocity (m/s)
			-23111049.826047	289.165842
2080	3400	2300	13056254.768326	494.145835
			28064.586272	3831.432914

The table 1 shows the CBERS-4 parameters. The first line, item a, shows the Two Line Element [a]. The mass of the satellite is set at 2080 kg with an Equivalent Isotropically Radiated Power – EIRP [b] of the satellite antenna of 3400 W. The antenna power [c] is 2300 W, and the components of the speed and position of the satellite CBERS-4, 24 hours after for the three coordinates is shown in the right column of the table.

The CBERS-4 satellite input parameters refer to March 14, 2016, at $11^h\ 14^m\ 15.23^s$, with the orbital elements, obtained using an executable program called *TrackStar Version 2.65,* which converts the ephemeris data in the format *TLS (Two Line Element)* (which contains the orbital elements), for CBERS-4 satellite state vector data (Table 1).

For the model used in this work, parameters such as antenna gain, satellite mass and total power radiated from the satellite antenna are important for the correct modeling of disturbances in the satellite orbit. The value of the angle of opening of the antenna is considered in the value of the maximum gain of the antenna of the satellite and, therefore, will be intrinsically considered in the calculations (figure 1).

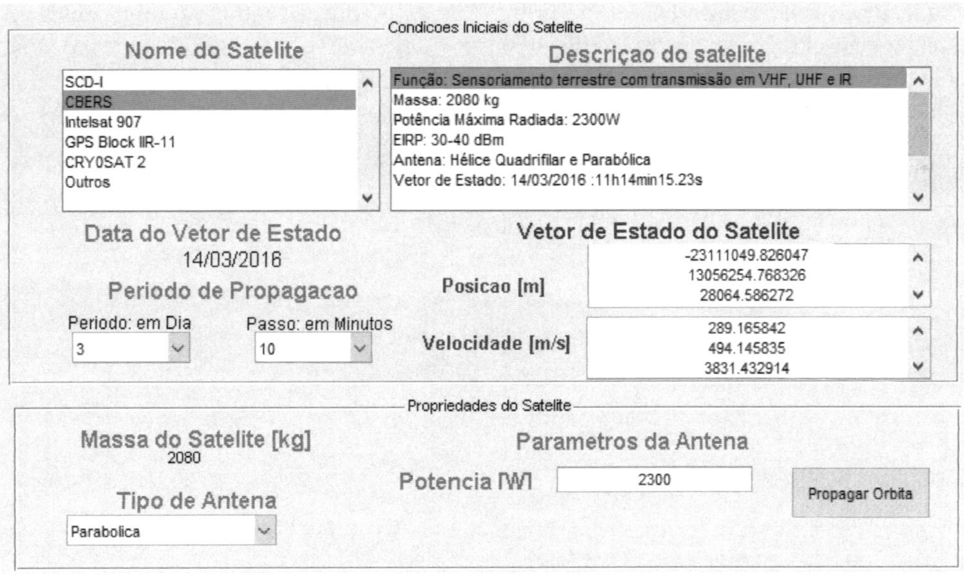

Figure 1. Parameter input screen of CBERS-4 satellite.

The input parameters are: satellite type, state vector, propagation period (in days and minutes), satellite mass in orbit, antenna type (helix quadrifilar or parabolic) and antenna power. After the inclusion of these parameters, the orbit was propagated by clicking on "Propagar Orbita". The result of the simulation is shown in the form of graphs as in Figures X.

As a result of the output of the program, we have a perspective of the orbit trajectory of the chosen satellite, and the possibility of constructing graphs that represent the behavior of its orbital elements.

3. Analysis of results

The orbit of the CBERS-4 satellite was propagated to a time interval of 5 days with step of 300 seconds, first with acceleration of the two bodies (reference orbit) and after this considering the Electromagnetic Reaction Acceleration Model (disturbed orbit). The results were then compared, between these orbits, by analyzing the positional deviations in the components radial (ΔR), normal (ΔN) and transversal (ΔT), according to the model.

Also was realized an analytic study of this disturbance on the orbital elements, especially on the semi-axis major, aiming to interpret its periodic behavior.

The calculation of propagation requires that are considered, for an electromagnetic disturbance, the initial state vector, satellite mass (kg), power radiated by the antenna (W) an the maximum antenna radiation gain (dB)

Considering the communication of the CBERS-4 from a parabolic antenna with 2300 W of power, it was observed that the velocity, semi-axis major, eccentricity, mean anomaly, longitude of ascending node, orbit slope, and argument of the perigee, did not suffer significant variations, being in the order of magnitude of 10^{-4} – 10^{-6}. However, admitting a parabolic antenna, the altitude of the CBERS-4 suffered a variation of approximately 920 m, the X-Y-Z coordinates presented variations of 730 cm, 400 cm and 620 cm, respectively. The components had values of 0,9 cm (Radial), 0,0 cm (Normal) and 1000 cm (Transversal) (figure 2). For a quadrifilar propeller antenna, with a EIRP of ~3400 W, the propagation of the CBERS-4, for the same 5 days (with step of 300 seconds), resulted in an altitude disturbance equal to the case of a parabolic antenna, that is, with a variation of 920 m. The X-Y-Z coordinates presented a disturbance of 220 cm, 110 cm and 180 cm, respectively, values with variations in magnitude lower than the disturbance case of the X-Y-Z coordinates with a parabolic antenna.

The components of direction presented disturbances of the order of: 0,9 cm (Radial), 0,0 (Normal) e 210 cm (Transversal), indicating that the case of the satellite use quadrifilar propeller antennas results in perturbations in the X-Y-Z and steering coordinates, lower than in the case of using parabolic type antennas (figure 3).

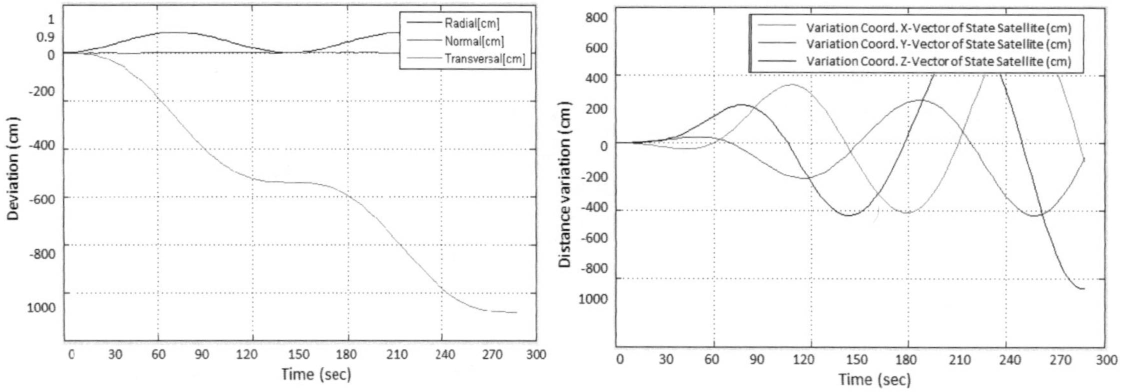

Figure 2. Disturbance in radial, normal and transversal components (left). Variation of the X-Y-Z coordinates of the CBERS-4 satellite, for 5 days (step of 300 seconds) (right), for a parabolic antenna.

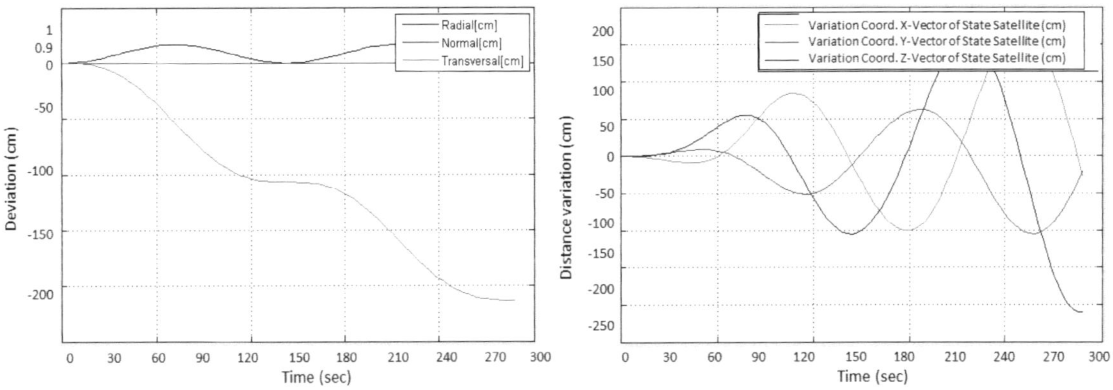

Figure 3. Disturbance in Radial, Normal and Transversal Components (left). Variation of the X-Y-Z coordinates of the CBERS-4 satellite, for (step of 300 seconds) (right), for a quadrifilar antenna.

The observed variations in the X-Y-Z coordinates and in the components of direction are the most disturbeds parameters, according Montenbruck and Gill (2005), Eshagh and Najafi (2007) and Solano et. al. (2012). It is necessary to establish corrections on these parameters, especially when dealing with a set of antennas, being these of the type parabolic and quadrifilar propeller. Due to the arrangement of antennas of this type infer a greater perturbation to the satellite, changing significantly the variables of the state vector and the components of direction.

4. Conclusion

With the orbital elements of the CBERS-4 satellite it was possible to obtain its corresponding parameters of the state vector, composed of Cartesian coordinates of position and velocity, in relation to the inertial reference system. The Electromagnetic Reaction Acceleration Model propagated its orbit from the March 14, 2016, at $11^h14^m15.23^s$, for 5 days and with step of 300 seconds, together with the mass value of this satellite, power radiated by the parabolic antenna (in Watt) and EIRP of the quadrifilar propeller antenna (in Watt). The effects of perturbation on the Radial, Normal and Transverse components of the CEBRS-4 satellite were simulated, both of its variations in the X-Y-Z coordinates, for two types of antennas: quadrifilar and parabolic propeller. It was observed that for both CBERS-4 satellite antennas, velocity, semi-major axis, orbit eccentricity, mean anomaly, longitude of ascending node, orbit slope, argument of perigee, did not suffer significant variations.

For the simulation with a parabolic antenna transmitting in the S-band/UHF frequency and its radiated power of 2300 W, the altitude, after the propagation of the orbit of the 5 days, obtained a variation of 920 m. The X-Y-Z coordinates had variations of 730 cm in X, 400 cm in Y and 620 cm in Z. For a quadrifilar propeller antenna, the variation in altitude was the same 920 m, while the coordinates X-Y-Z had variations of 220 cm in X, 110 cm in Y e 180 cm in Z.

The simulations also showed that for a parabolic antenna operating in this frequency range, it implies a variation of 0.9 cm in the Radial direction, 0,0 cm in the Normal direction and 1000 cm in the Transversal direction. Meanwhile, in the case of a quadrifilar propeller antenna, the component variations were 0.9 cm in the Radial direction, 0.0 in the Normal direction and 210 cm in the Transversal direction.

Although the CBERS-4 satellite has antennas of both types the results show that, from this simulation, the satellite communication operations with a terrestrial antenna of the quadrifilar propeller type offers on this satellite a lesser electromagnetic disturbance, when compared to a parabolic antenna, even the quadrifilar propeller antenna having an equivalent isotropically radiated power de 3400 W, compared to the radiation power of the parabolic antenna, of 2300 W. These simulations considered only the propagation of the CBERS-4 satellite orbit from the electromagnetic disturbance of one antenna

IOP Conf. Series: Journal of Physics: Conf. Series **911** (2017) 012003 doi:10.1088/1742-6596/911/1/012003

at a time, when in reality the antennas operate in an arrangement, which should increase the perturbation values at the X-Y-Z coordinates, and the Radial, Normal and Transversal components.

References

[1] Bukhsh W A and Jonsson B L G 2010 Element Position Perturbation for a Narrow Spot Beam with Applications to Satellite Communication Antennas *Progress in Electromagnetics Research* PIER **104** pp 283- 295

[2] Cook G E 1962 Luni-solar Perturbations of the Orbit of an Earth Satellite. *Geophysical Journal of the Royal Astronomical Society* vol 6 issue 3 pp 271–291

[3] Cojocaru S 2007 A Numerical Approach to GPS Satellite Perturbed Orbit Computation *The Journal of Navigation* **60** pp 483–495

[4] Eshagh M Najafi Alamdari M 2007 *Perturbations in Orbital Elements of a Low Earth Orbiting Satellite* Geodesy Department

[5] Heilmann A, Ferreira L D D and Dartora C A 2012 Perturbative Effects of Antenna Radiation Reaction on Artificial Satellite *Aerospace Science and Technology* **23** pp 352 – 357

[6] Heilmann A, Ferreira L D D and Dartora C A 2013 Antenna Radiation Effects on the Orbits of GPS and INTELSAT satellites *Acta Astronautica* **88** pp 1 – 7

[7] Maral G and Bousquet M 1998 *Satellite Communications Systems: Systems, Techniques and Technology* 3nd ed England John Wiley & Sons Ltd

[8] Montenbruck O and Gill E 2005 *Satellite Orbits: Models, Methods and Applications* Springer

[9] Seeber G 1993 *Satellite Geodesy: Foundations, Methods and Aplications* Berlim W Gruyter

[10] Solano C J R, Hugentobler U and Steigernberg P 2012 Impact of Albedo Radiation on GPS Satellites *Institute for Astronomical and Physical Geodesy* Munich Germany

[11] Ziebart M, Sibthorpe A and Cross P 2007 Cracking the GPS – SLR Orbit Anomaly *ION GNSS 20th International Technical Meeting of the Satellite Division* Fort Worth Texas September pp 25-28

[12] Zhou B, Liu Q and He X 2008 Research on the Electromagnetic Interference of Antennas on the Satellite *Piers online* **4** no 3

Gamma radiation in ceramic capacitors: a study for space missions

Eduardo dos Santos Ferreira[1], **Juliana Sarango Souza**[1]

[1]CECS, Federal University of ABC, São Paulo Avenue s/n, São Bernardo do Campo, Sao Paulo, BR.

E-mail: eduardo.ferreira@ufabc.edu.br

Abstract. We studied the real time effects of the gamma radiation in ceramic capacitors, in order to evaluate the effects of cosmic radiation on these devices. Space missions have electronic circuits with various types of devices, many studies have been done on semiconductor devices exposed to gamma radiation, but almost no studies for passive components, in particular ceramic capacitors. Commercially sold ceramic capacitors were exposed to gamma radiation, and the capacitance was measured before and after exposure. The results clearly show that the capacitance decreases with exposure to gamma radiation. We confirmed this observation in a real time capacitance measurement, obtained using a data logging system developed by us using the open source Arduino platform.

1. Introduction

An embedded system is an electronic circuit coupled to a complex mechanical system. This electronic circuit can be analog or digital, and it has the function of making the control of these systems, or health monitoring of complex electromechanical systems.

The cosmic radiation causes changes the electronic circuits behavior, such as space missions are heavily exposed to these radiations they need proper shield to their electronic systems. Long operating range missions require great precision in inertial navigation electronic systems. The shield do not fully prevent the radioactive penetration, as well passive components end up accumulating doses during the long stay in space, and therefore undergo changes in its operation.

Capacitors are part of embedded systems, and their behavior is directly related to its dielectric, which is the insulating material. The dielectric has the function of facilitating the flow of the electric field and the dielectric constant is the parameter that measures this ability. A ceramic capacitor has a high dielectric constant, and insulating material is used as a mixture of quartz, feldspar and alumina. Each of these minerals have a significant contribution to the electrical properties of the capacitor.

When the ceramic insulator is irradiated by some kind of ionizing radiation, electrons are raised to the conduction band, leaving holes at the valence band. The pre-existence of defects inside of insulator can trap these free charges, changing the amount of free carriers and electric dipoles in the material. Ionizing radiation causes an increase in the number of free carriers, and a reduction in the number of electric dipoles, which in turn reduces the electrical permeability of the alumina and hence reduces the capacitance of the material [1]. Subsequent stimulation by heat or light leads to absorption of energy by the trapped electrons, causing a transition to the conduction band and posterior recombination with localized trapped holes resulting in luminescence emission [2].

The ASTER mission is the first attempt of Brazil to send an object to deep space [3]. The mission will have navigation and attitude control systems [4] and measuring instruments such as a topographic profile system [5].

This study sought to determine the changes in capacitance of ceramic capacitors due to exposure to doses between 1 and 20 Gy. These capacitors have alumina as dielectric material, and observed in earlier studies the capacitance of alumina decreases with increasing dose, due to the ionization processes and the disruptions of electric dipoles within the material [1].

Therefore, in this study we observed the effects of gamma radiation in ceramic capacitors, measuring the capacitance before and after exposure. We study the permanence of the effects in long time, if there is extinction effect after heating, as it occurs in the alumina thermoluminescence processes [6]. In addition, we made a real time measurement of capacitance during exposure to gamma source, using a data logging system dedicated to this purpose and developed by our group using the open source Arduino platform.

2. Experimental

We use ceramic capacitors obtained in the retail of electronic components, the nominal values are 10 nF and 100 nF, 50 V isolation voltage and tolerance of 20% in the nominal value. Each capacitor was identified and their capacitance was measured using precision capacimeter MINIPA MXB-821.

We recorded the actual values of each capacitor and then sent to irradiate in samples with three capacitors. Each sample was irradiated with a specific dose, and the values range was 1 to 20 Gy. We used a ^{60}Co gamma-source in RT, with 28.76 Gy/h. After exposure, we measured capacitance values and calculate the change in capacitance:

$$\Delta C = C_a - C_b \tag{1}$$

where C_a is the average capacitance before exposure and C_b is the average capacitance after.

To determine if the effect of radiation is permanent and there is a relation with the thermoluminescence phenomena, we repeated the procedure with intervals of 7 and 37 days. In addition, we repeat the procedure and then heat it to 300 °C for 30 s and then measure the capacitance values.

Finally, we build an autonomous system of capacitance measurement, in order to record the effect of radiation on the capacitor in real time. The capacitance measurement consisted of measuring the capacitor charge time tau in a series resistor capacitor circuit:

$$C = \frac{1}{R \cdot \tau} \tag{2}$$

where C is the instantaneous capacitance, R is the series resistor (1 MΩ), and τ is the time at which the capacitor voltage reaches 63.2% of the applied voltage in the circuit [7].

An Arduino microcontroller was used to perform the measurement time constant. The RC circuit is connected to a microcontroller port, and the capacitor is connected to the input of the analog to digital converter (AD). A third pin is used to completely discharge the capacitor. The Arduino input-output pins (IOpins) can be in one of these tree electrical states, a high output state (5 V), low output state (0 V) and a high impedance state (tristate). A high output state and low output state have a low impedance, because the IOpins is in output mode. When the IOpin is in input state, it has a high input impedance. The complete experimental apparatus to capacitance measurement is shown in figure 1. The capacitance measurement algorithm is shown in table 1.

IOP Conf. Series: Journal of Physics: Conf. Series **911** (2017) 012004 doi:10.1088/1742-6596/911/1/012004

Table 1. Capacitance measurement algorithm

Step	Action
1	Set discharge pin to input mode.
2	Set charge pin to output mode.
3	Get the start time $t_s = t$ in the time register.
4	Set charge pin to high output.
5	Check the voltage in A/D register.
6	Wait until the voltage on the A/D register is equal to 63.2% of output voltage.
7	Calculate the charge time: $\tau = t_i - t$.
8	Write τ value in a small disk card (SD card).
9	Set discharge pin to output mode.
10	Set discharge pin to low output.
11	Go to step 1 until it ends.

Additionally it was placed a charge counter circuit to measure the emission source during the experiment. The sensor is a PIN photo-diode, model TEND5000, covered with aluminum foil to filter the alpha and beta particles. An amplifier circuit charge counter in conjunction with a voltage comparator is used to generate pulses for each received gamma photon. These pulses are connected directly to the clock pin of the Arduino counters. The main program will stop counting at scheduled time cycles for each 30 s, calculates how many pulses were counted during 1 s, and stores the value of counts per second on a memory card.

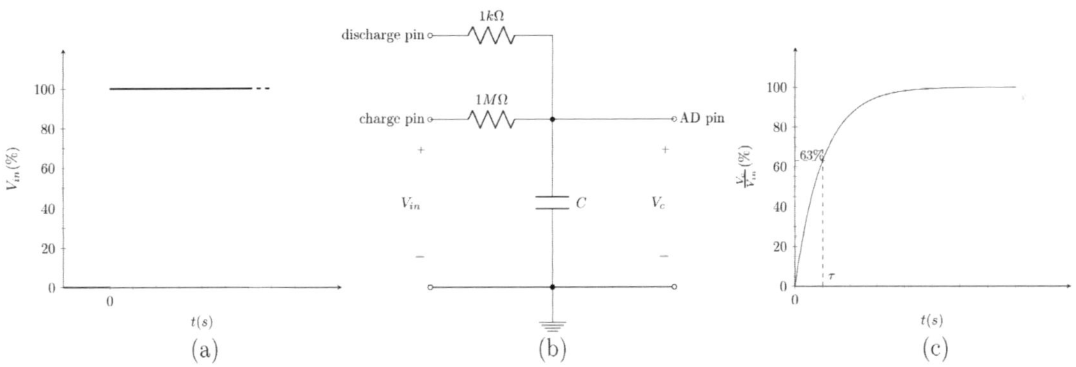

Figure 1. Capacitance measurement setup: (a) input test voltage, (b) capacitor under test circuit and (c) Capacitor charge behavior.

3. Results

Figure 2 shows the results of changing capacitance as a function of absorbed dose in 10 nF capacitors. It is observed that the higher the dose, the greater the difference in capacitance, agreeing with the results previously presented in our studies [1]. To confirm the results, repeat the test with 100 nF capacitor using an Agilent 4294A impedance analyzer, figure 3 shows

IOP Conf. Series: Journal of Physics: Conf. Series **911** (2017) 012004 doi:10.1088/1742-6596/911/1/012004

the results. The curves of Figures 2 and 3 have a similar behavior but the magnitude ΔC is greater in figure 3. This difference is due to difference of calibration equipment and the type of meter, in figure 2 is a capacimeter, while equipment in figure 3 is an impedance analyzer. The impedance analyzer is an instrument that measures the complex impedance of the device, and from this information it calculates the capacitance. The capacimeter measuring capacitance directly, through the charge and discharge method.

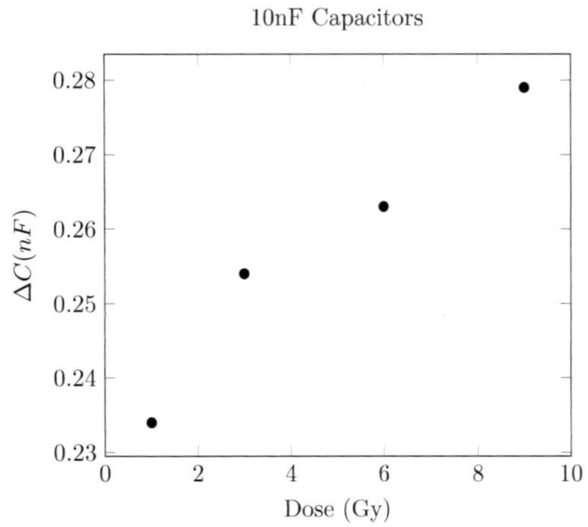

Figure 2. Changing in capacitance due gamma radiation for 10 nF capacitors.

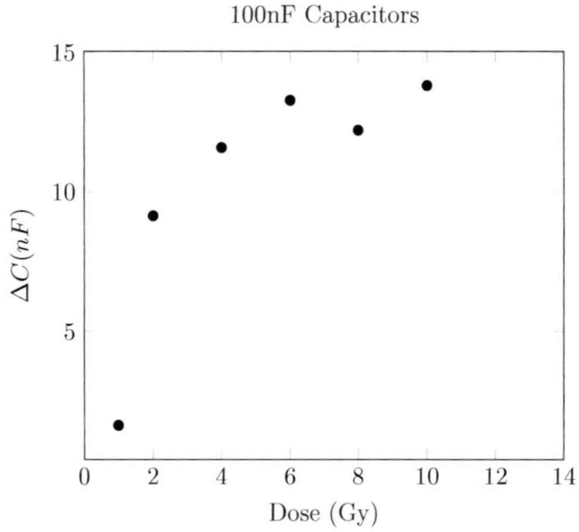

Figure 3. Changing in capacitance due gamma radiation for 100nF capacitors.

Figure 4 shows the result of the capacitance change to a 100 nF capacitor during exposure to gamma rays. In the same figure, we show the emission intensity of the source in counts per second (CPS). After the experiment the component absorbed approximately 19.6 Gy. As expected, we found that the device capacitance decreases during exposure. The total capacitance decrease was approximately 3 nF.

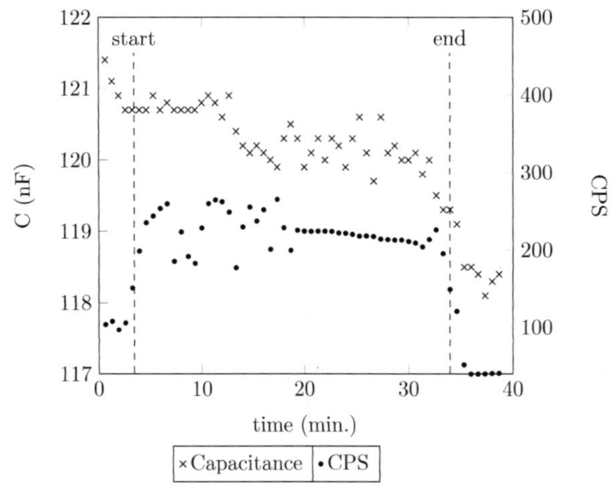

Figure 4. Changing in capacitance due gamma radiation for 100nF capacitors.

To determine the effect of radiation on the capacitor is permanent and if it is correlated with the thermo-luminescence (TL) phenomena, repeat the test and follow the deltaC variation due to the dose over time and with heat treatment. The results are displayed in table 2. We note that the effect tends to disappear with the passing of time, indicating that the phenomenon is unstable, probably the ink covering the capacitors is not enough to contain all the ambient light, and should cause a decay as is seen in photoemission ceramic materials in optically stimulated luminescence measurements (OSL). The heating of the capacitors eliminates all the changes caused by radiation. Thus, we believe that the observed effect is related to processes involved in thermoluminescence (TL).

Table 2. Capacitance permanence test

	ΔC (nF)		
Dose (Gy)	7 days	37 days	300 °C for 30 s
1	0.23	0.05	0
3	0.25	0.09	0
6	0.28	0.11	0
9	0.30	0.22	0
20	0.34	0.2	0

4. Conclusions

From an engineering point of view, the change in the observed capacitance is significant, therefore, a good shielding of electronic circuitry is required for deep space missions, or from the results obtained, we can create compensation circuits for the dose absorbed by the systems electronics, eliminating the shield and gaining space for more payload on the mission.

Gamma radiation incident on the ceramic capacitors was not enough to make permanent changes occur in the dielectric, justifying the return of the original value of capacitance within a week. For be temporary, the capacitor can be used to detect gamma radiation. Or, it could be calculated by measuring the absorbed dose reduction of capacitance in the first 24 hours in the event of a nuclear accident.

5. Acknowledgments

We thank the Institute of Nuclear Energy Research (IPEN-SP) for offering us the use the gamma source, which was essential for the development and completion of this work.

References

[1] Ferreira E S, Gonçalves K A, Mittani J C R, Yee M and Tatumi S H 2014 Gamma radiation effects in capacitance of alpha alumina doped with rare earths *Radiation physics and chemistry* **95** 385
[2] McKeever S W S 1998 *Thermoluminescence of solids* (New York: Cambridge University Press)
[3] Sukhanov A A, Velho H F C, Macau E E and Winter O C 2010 The Aster project flight to a near-earth asteroid *Cosmic Research* **48** 443
[4] Sarli B V, Silva A L and Paglione P 2013 Preliminary attitude control studies for the ASTER mission *Journal of physics conference series* **465** 1742
[5] Brum A G V and Rodrigues 2013 A P Topographic profile of a target with use of laser pulses: a survey directed to the Brazilian deep space mission ASTER *Journal of physics conference series* **465** 1742
[6] Gonçalves K A, Bitencourt J F S, Mittani J C R and Tatumi S H 2010 Study of radiation induced effects in the luminescence of nanostructured $Al_2O_3 : Yb$, Er crystals *IOP Conf. Ser. Mater. Sci. Eng.* **15** 012080
[7] Hayt W H, J. E. Kemmerly and Durbin S M 2006 *Engineering Circuit Analysis* (New York: McGraw-Hill Book Company Inc)

IOP Conf. Series: Journal of Physics: Conf. Series **911** (2017) 012005 doi:10.1088/1742-6596/911/1/012005

Analysis of the passage of a spacecraft between the Van Allen belts considering a low and high solar activity

M R da Silva[1] **and E M Rocco**[1]

[1]Space Mechanics and Control Division (DMC), National Institute for Space Research (INPE), São José dos Campos, São Paulo, Brazil

E-mail: doc.silvamr@gmail.com, evandro.rocco@inpe.br

Abstract. The radiation in the Van Allen belts, produced by electrically charged particles, can cause damages to the electrical equipments of a satellite in orbit of Earth. In this paper, the Van Allen belts are modeled using the data from the space mission Van Allen Probes Mission. With this model, a study was made taking into account the passage of a spacecraft through the Van Allen belts estimating the absorbed radiation dose and the time that spacecraft remained in the radiation zones, considering the effects of a low and high solar activity.

1. Introduction

In 1958, the satellites Explorer I and Explorer III were launched with the goal to collect data about space environment around the Earth and both confirmed the existence of a charged particles ring [1, 2]. The existence of a ring of particles was accept and included the idea that these particles would be trapped in this ring. However, the main attention these experiments was to the nature this radiation belt. According to [3], there are two radiation belts and they are formed by protons and electrons trapped in the geomagnetic field but their formation differ each other. The inner belt contains more protons than the outer belt, which contains more electrons. Furthermore, between the belts, there is a region where the intensity of radiation is weak but it can to vary during a time as the outer belt owing to solar activity [4].

Nevertheless, a new structure of radiation was discovered. In 2012, NASA launched to the space the twins spacecrafts, named Radiation Belt Storm Probes (RBSP), which are part of Van Allen Probes Mission [5]. In a experiment realized in the period of September to October in 2012, the probes discovered a third structure of radiation or a third belt between the inner and outer belts [6]. The appearance of third radiation zone is associated with the passage of solar winds, which formed waves of ultra-low frequency with large amplitudes, responsible for the loss of low energy electrons in the external radiation belt [7].

The importance in studying these belts is because the electronic components on board are sensitive to space radiation dose. In this paper, the main objective is modeling the Van Allen belts using the data provided by Van Allen Probes Mission during the months of September to November 2012 and January to March 2015. From this model, will be simulated the impulsive and low-thrust orbital maneuvers and the combination between them in the periods mentioned previously. For each of these maneuvers, will be determined the time of passage of the spacecraft

IOP Conf. Series: Journal of Physics: Conf. Series **911** (2017) 012005 doi:10.1088/1742-6596/911/1/012005

through the regions of radiation and the radiation dose accumulated in the spacecraft considering silicon, aluminum, lead, tantalum and tungsten as shielding materials.

2. Modeling the Radiation Belts

In the work of [8], the radiation belts were modeled, as a first approach, considering just the geographical latitude angle and the distance that separate the radiation zones. However, it is known that the radiation belts are not stable and vary in its sizes owing to solar activity. In this paper, the Van Allen radiation belts were modeled using the data provided by the Van Allen Probes Mission. This mission began at the end of August 2012 and is currently active with the objective of studying the dynamics of radiation belts using the twins probes, RBSP-A and RBSP-B, whose orbits are geocentric, elliptical and with a inclination of 10° [5].

However, during Van Allen Probes Mission data analysis, it was found that space probes can not fully map the radiation belts in a single day, which leads to the appearance of failures in filling flux intensity values for these regions. In this paper, to homogenize the regions of radiation, according to their delimitation, it was applied a method of interpolation of the values of radiation flux.

This method consists in:

(i) Obtain the angles of geographical latitude and longitude from the positions of the probes, in the ECI reference;

(ii) It is considered that the Van Allen radiation belts would be modeled on the plane of the magnetic equator. So the angles of geographical latitude and longitude were transformed to the angles of magnetic latitude and longitude;

(iii) In addition, it is considered that during a period of one day, the radiation belts do not vary, that is, the Van Allen belt has the same characteristic for all values of the magnetic longitude.

(iv) Therefore, the interpolation method will only require the radiation flux values as a function of the magnetic latitude.

3. Dose of Radiation

In this work, the absorbed radiation dose equivalent rate is determined from the definitions of radiation dose and omnidirectional radiation flux, so that

$$\dot{D}(E) = w_R \kappa \frac{\Phi(E)E^2}{\rho t} \tag{1}$$

where w_R is the radiation weighting factor, such that 1 is for electron and 2 is for proton, κ is a constant equal to $4.80653199.10^{-11}$, Φ is the omnidirectional flux, in function of the energy E, r_S is the radius of spacecraft, t is the shielding thickness and ρ is the density of shielding material. The unit of absorbed radiation dose equivalent rate is rad/s, where rad is a unit of absorbed radiation dose defined as 1 rad = 0.01J/kg [9].

4. About the Simulations

The simulations were performed in the STRS (Spacecraft Trajectory Simulator) [10, 11, 12, 13]. The Van Allen belts are delimited by McIlwain's parameter [14], omnidirectional flux of particles (electrons and protons) and the magnetic latitude, where the inner belt is bounded between ±40° and the outer belt is bounded between ±65°.

The spacecraft will start from a low orbit around the Earth and will pass through the radiation regions. For this, we consider the low-thrust and impulsive orbital maneuvers and the combination between them, seeking for the minimum time that the spacecraft can remain in the regions of radiation. So, the accumulated radiation dose equivalent rate will be estimated using the Eq.1, such that the spacecraft is considered as a perfect sphere with a radius of 1 m and aluminum, tungsten, lead, tantalum and silicon are the shielding materials for a layer of 1 mm of thickness. The values of densities of the shielding materials used are shown in the Tab.1.

Table 1. Densities of materials shielding.

Materials	Density (kg/cm^3)
Aluminum	0.002697
Tungsten	0.019250
Lead	0.011340
Tantalum	0.016650
Silicon	0.002330

Furthermore, it is considered the effects of external perturbations as gravitational attraction of the Sun, solar radiation pressure and the terms J_2, J_3, J_4, J_5 and J_6 of the Earth's and Moon's gravitational potentials. The control of the spacecraft's trajectory is performed by a PID controller in a closed-loop system.

5. Results

In the period of September 2012, it was observed and expected the presence of a third structure of radiation, for the region of electrons with energy equal to 4.2 MeV, because of the high solar activity (see Fig. 1)[6]. However, during the period of January 2015, during a low solar activity, only the presence of the classical zones of radiation (outer and inner belt) was observed, as shown in the Fig. 2. In those figures, the directions X and Z are defined by magnetic latitude and longitude, where ER is the Earth radius.

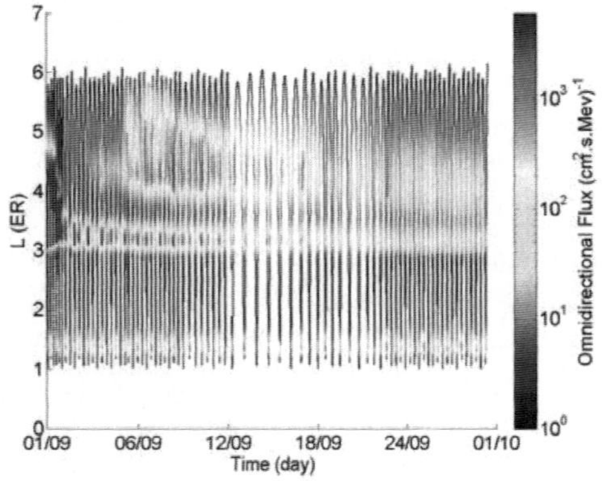

Figure 1. Electron radiation zones in September 2012 (E = 4.2 MeV).

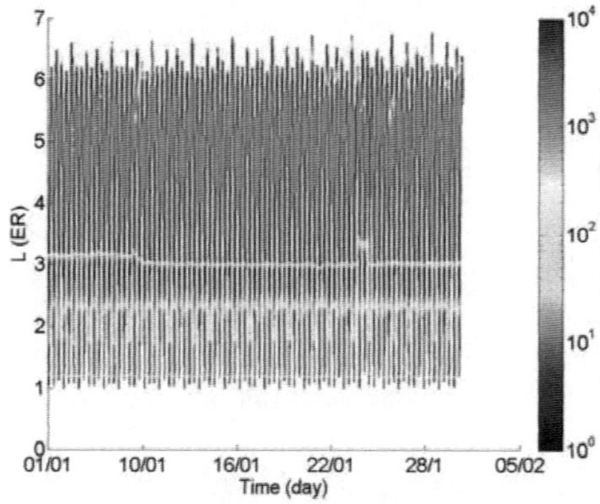

Figure 2. Electron radiation zones in January 2015 (E = 1.8 MeV).

For regions that contain only protons, no type of anomaly appeared, regardless of the variation of solar activity. Thus, in this work, we simulated the passage of a spacecraft through regions of radiation containing only electrons with energy equal to 4.2 MeV (September-November 2012) and 1.8 MeV (January-March 2015), and for regions of radiation containing only protons with energy equal to 21.25 MeV.

The Fig.3-4 are examples of the results obtained by simulating the spacecraft passage through the Van Allen radiation belts using Hohmann maneuvers and its absorbed radiation dose equivalent rate, respectively. In the Fig. 3, the magenta color line represents the trajectory of spacecraft and the lines with hot and cold colors, forming the map of the region of radiation, represent the omnidirectional flux of electrically charged particles based on the passage of the RBSP space probes through the radiation regions, in a certain period of time that remained in the space.

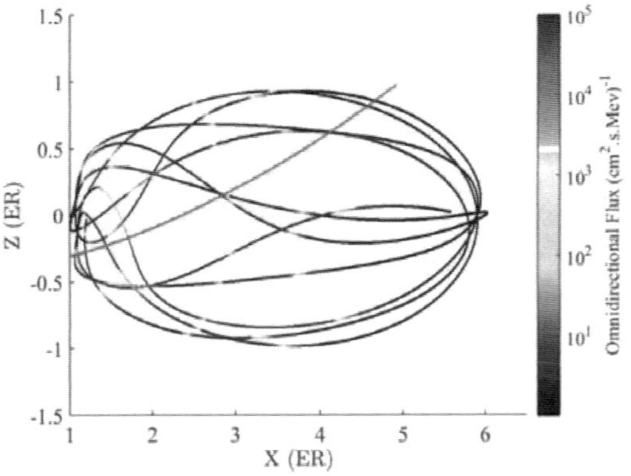

Figure 3. Passage of the spacecraft through the region of electrons with energy equal to 4.2 MeV using Hohmann maneuver in September 2012.

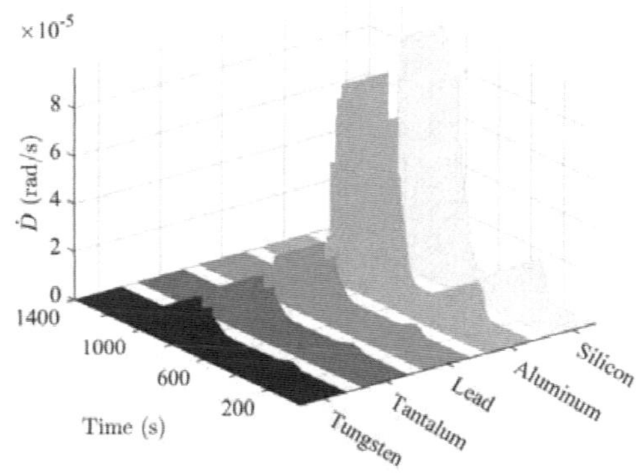

Figure 4. Absorbed radiation dose equivalent rate for a region of electrons with energy equal to 4.2 MeV in September 2012.

The Fig.5-8 show the results of the spacecraft passage through the Van Allen radiation belts using low-thrust maneuvers and their absorbed radiation dose equivalent rate, respectively. In the Fig.5, it is observed, initially, that the spacecraft remained only in the region of the inner belt and continues further in the region as the semi-major axis of the orbit increases until the spacecraft is injected out of the region of radiation, as shown in the Fig.6.

The results of the combination of Hohmann with the low-thrust maneuver are shown in Fig.9-10. Initially, the Hohmann maneuver is applied so that the spacecraft crosses the inner belt and is then injected into the slot where it remains by means of the low thrust maneuver. Upon approaching the region of the outer radiation zone, the spacecraft again uses the Hohmann maneuver, being injected out of the radiation region where it continues, again with the low-thrust maneuver.

The Tab.2 show the time of passage and the maximum radiation dosage rate accumulated in the shielding layers of tantalum, tungsten, and lead in the period of September-November 2012 and January-March 2015.

6. Conclusions

This study aimed to determine the time in which the spacecraft remained in the regions of radiation, around the Earth. Firstly, the electron and proton radiation regions were modeled with Van Allen Probes Mission data using the interpolation method, as described in the Sec.2. Through this, it became possible to verify, in the period of high solar activity, the presence of a new radiation structure, known as the third Van Allen belt formed between the inner and

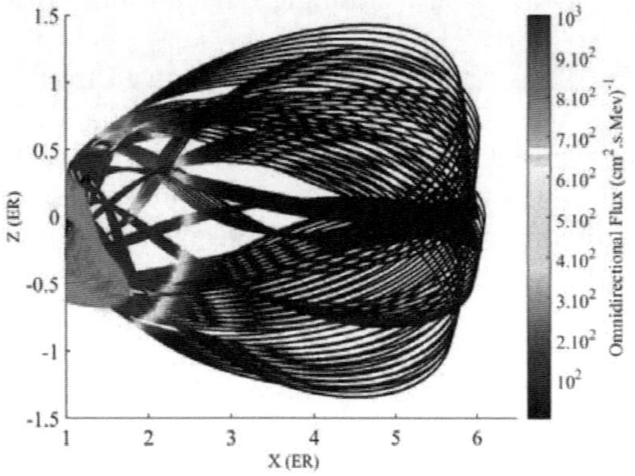

Figure 5. Passage of the spacecraft through the region of protons with energy equal to 21.25 MeV using the low-thrust maneuver in September 2012.

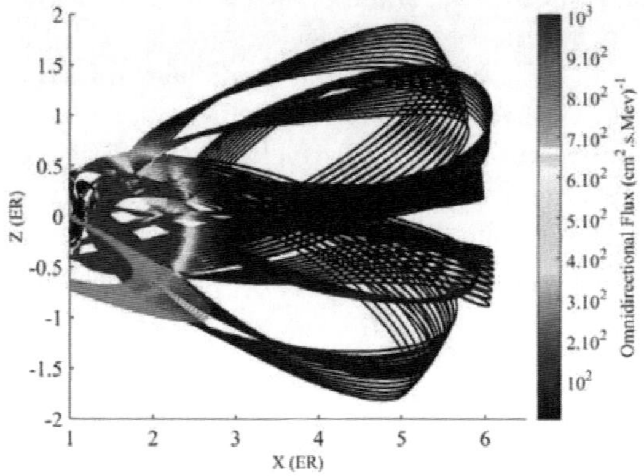

Figure 6. Passage of the spacecraft through the region of protons with energy equal to 21.25 MeV using the low-thrust maneuver in October 2012.

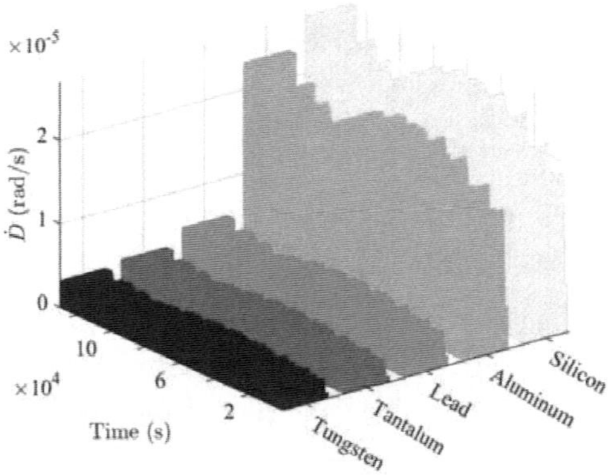

Figure 7. Absorbed radiation dose equivalent rate for a region of protons with energy equal to 21.25 MeV in September 2012.

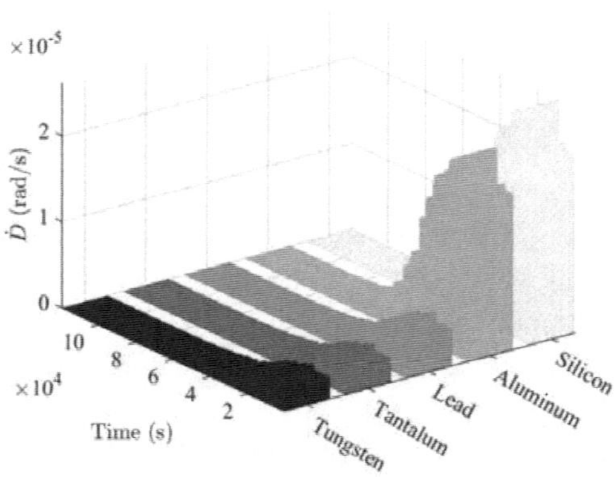

Figure 8. Absorbed radiation dose equivalent rate for a region of protons with energy equal to 21.25 MeV in October 2012.

Table 2. Time of passage of the spacecraft through the radiation zones (days) and absorbed radiation dose equivalent rate (rad/s) between parentheses.

	2012		2015	
	Electron	Proton	Electron	Proton
H	$< 1 - (1.10^{-5})$	$< 1 - (5.10^{-6})$	$< 1 - (5.10^{-3})$	$< 1 - (5.10^{-6})$
L-T	$50 - (1.10^{-4})$	$35 - (5.10^{-6})$	$58 - (5.10^{-3})$	$34 - (5.10^{-6})$
H + L-T	$14 - (3.10^{-5})$		$2 - (2.10^{-3})$	

outer belts. This new structure was observed for a region of electrons with energy equal to 4.2 MeV in the period of September 2012. In the region of radiation that contains only protons,

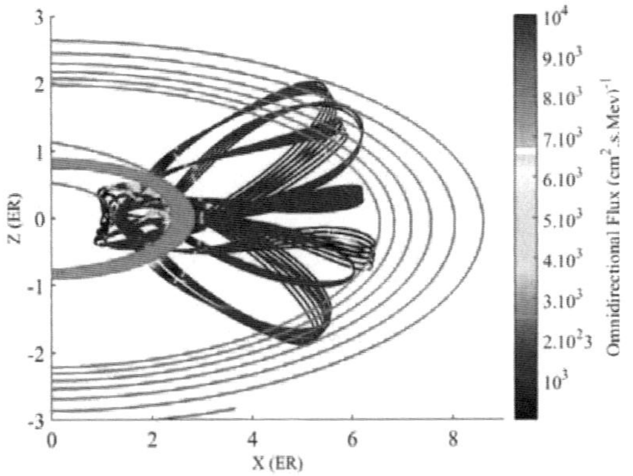

Figure 9. Passage of the spacecraft through the region of electrons with energy equal to 1.8 MeV using the Hohmann and low-thrust maneuver in January 2015.

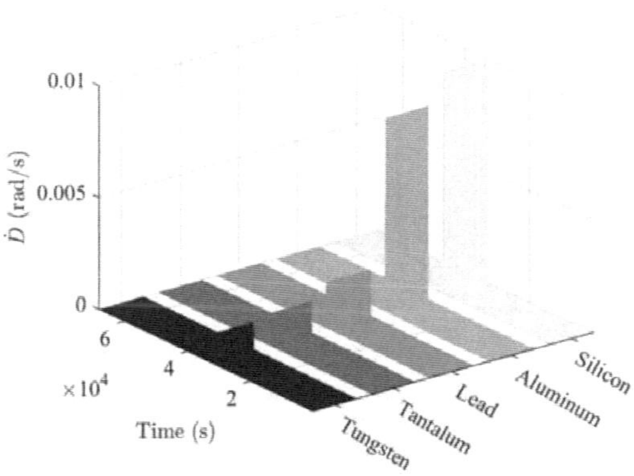

Figure 10. Absorbed radiation dose equivalent rate for a region of electrons with energy equal to 1.8 MeV in January 2015.

it did not observe any type of anomaly. In January 2015, only the presence of the classical radiation structures (inner and outer belts), during the low solar activity, was verified for a region of electrons with energy equal to 1.8 MeV. Thus, Hohmann maneuver, low-thrust and the combination between them were simulated for regions containing only electrons and protons. For all the cases studied, it was observed that the Hohmann maneuver was the only one that managed to keep the spacecraft in minimum time in the radiation belts, while that the low-thrust maneuver maintained it in the longest time, as expected. The absorbed radiation dose equivalent rate was calculated and it was found that lead, tantalum and tungsten were the materials that received least radiation in their shielding layers, regardless of periods of solar activity. For the proton regions, the absorbed radiation dose equivalent rate were the same during Hohmann and low-thrust maneuvers. However, for the electron regions, the Hohmann maneuver was the only one that obtained the lowest rate in the period of high solar activity, whereas the combination between the Hohmann maneuvers and low-thrust obtained the lowest rate in the period of low solar activity .

References

[1] Van Allen J A, Ludwig G H, Ray E C and Mcilwain C E 1958 *J. Jet Prop.* **28** 9 588–92
[2] Rohrlich F and Van Allen J A 1960 *First Public Lecture on the Discovery of the Geomagnetically Trapped Radiation* (Department of Physics and Astronomy: State University of Iowa) p 60
[3] Van Allen J A 1959 *J. Geophys. Res.* **64** 1683–89
[4] Parker N E 1960 *J. Geophys. Res.* **67** 10 3117–30
[5] Kessel R L, Fox N J and Weiss M 2012 *Space Sci. Rev. J.* **179** 1 531–43
[6] Baker D N *et al.* 2013 *Science* **340** 186 186–90
[7] Mann I R et al 2016 *Nat. Phys.* **12** 978–83
[8] Oliveira T C, Rocco E M, Prado A F B and Ferreira J L A 2013 *J. Phys.: Conf. Series* **465**
[9] Haffner J W 1967 *Radiation and shielding in space* (New York and London: Academic Press) chapter IV p 112
[10] Rocco E M 2008 *XIV Coloquio Brasileiro de Dinamica Orbital* (Aguas de Lindoia) 17–21
[11] Rocco E M 2012 *Congresso Nacional de Engenharia Mecancia* (Maranhao)
[12] Rocco E M 2013 *J. Phys.: Conf. Series* **465**
[13] Rocco E M 2015 *J. Phys.: Conf. Series* **641**
[14] McIlwain C E 1961 *J. Geophys. Res.* **66** 11 3681–91

IOP Conf. Series: Journal of Physics: Conf. Series **911** (2017) 012006 doi:10.1088/1742-6596/911/1/012006

Solar Power Satellite system in formation on a common geostationary orbit

F J T Salazar and O C Winter

Grupo de Dinâmica Orbital & Planetologia, Univ. Estadual Paulista, UNESP, Guaratinguetá, Brazil

E-mail: e7940@hotmail.com, ocwinter@gmail.com

Abstract. The diurnal day-night cycle severely limits the Terrestrial solar power. To overcome this limitation, a Solar Power Satellite (SPS) system, consisting of a sunlight reflector and a microwave energy generator-transmitter in formation, is presented in this work. The microwave transmitting satellite (MTS) is placed on a common geostationary orbit (GEO) in the Earth's equatorial plane, and the sunlight reflector uses the solar radiation pressure to achieve quasi-periodic orbits about the MTS, so that the sunlight is always redirected to the MTS, which converts the solar energy in electromagnetic power and transmits it by microwaves to an Earth-receiving antenna. Assuming the sun line direction constant at different seasons (i.e. autumn/spring equinoxes and winter and summer solstices), previous studies have shown the existence of a family of displaced ecliptic orbits above or below the equatorial plane of the Earth around a GEO. In this study, the position of the Sun is assumed on the ecliptic plane with a mean obliquity (inclination of Earth's equator with respect to the ecliptic) of 23.5°. A linear solution as an initial condition for the full equations of motions about a GEO, which yields bounded orbit for the sunlight reflector about the MTS in the Earth-satellite two-body problem with solar radiation pressure. To redirect the sunlight to the MTS, the law of reflection is satisfied by the space mirror attitude.

1. Introduction

In the last decades, alternative energy sources have been proposed to minimize the greenhouse gases effects. Among these, solar energy is the first energy source in the world. However, any solar energy collection system is severely limited by the diurnal day-night cycle. In this manner, previous studies have proposed a solar-power system, consisting of a single large light-weight solar reflector (or a constellation of small solar reflectors), that collects solar energy, even at night, orbiting on a low Earth orbit (LEO) or a geostationary orbit (GEO), instead of collecting solar power by using solar panels on Earth's surface [1], [2], [3], [4]. Note that this kind of space system implies large amounts of solar energy being transported from orbit thousands of miles above Earth's surface. To minimize the problems with energy transportation, a satellite formation flying formed by a sunlight reflector and a microwave transmitting satellite (MTS) on a common GEO, has been presented by several authors [5], [6], [7] [8]. Thus, this Solar Power Satellite (SPS) system would use a space mirror to reflect the solar power energy to an energy generator-transmitter (collector) placed on a nominal geostationary orbit, which would transmit it to any Earth-receiving antenna, reducing the problems with energy transportation.

Takeichi et al. [9] present an Earth-pointing MTS on a common GEO and Sun-pointing reflectors that use the component of sunlight reflector acceleration perpendicular to the Earth's equatorial plane to levitate few kilometers above or below the equatorial plane and achieve orbits parallel to the GEO and have the same radius. However, due to the component of sunlight reflector acceleration parallel to the Earths equatorial plane is not enough to maintain such light levitation [10], [11], the SPS system proposed by Takeichi et al. needs a continuous orbital control by thrusters to maintain the longitude of the system [9] [12]. In this manner, Baig and McInnes [13] showed recently that, using this parallel component of reflector acceleration, it is possible to obtain a family of displaced ecliptic orbits for space mirrors, parallel to the equatorial plane, around a GEO point, i.e. periodic orbits with respect to an Earth-fixed rotating frame at a GEO point. Although, Takeichi et al. [9] and Baig and McInnes [13] show that displaced orbits are feasible, these results are only valid for a few days motion since the sun-line direction is assumed constant at different seasons (equinoxes and solstices). This approximation also permits to find analytical solutions for the linear equations of motion around a GEO point [13], and maintain the sunlight reflector attitude constant along the displaced ecliptic solution.

In this paper, the position of Sun is considered on the ecliptic (Earth's orbit plane), assuming a a mean obliquity equal to 23.5°, and the MTS on a GEO point. In this more realistic model, only numerical solutions exists, and the space mirror orientation satisfies the law of reflection to redirect the sunlight to the MTS along the solutions, i.e. the sunlight reflector pitch angle is no longer constant. Thus, this work uses the linear approximation found by Baig and McInnes [13] as an initial condition for the full equations of motion of the sunlight reflector about the GEO point in the Earth-satellite two-body problem, taking into account the effects solar radiation pressure. Bounded orbits for the SPS system are computed.

2. Equations of motion

A MTS on a common GEO moves around the Earth in a high circular orbit (42,164.1696 km measured from the center of the Earth), directly above the equator with a zero angle of inclination, and with an orbital period τ_e equal to the Earth's rotation period, i.e. $\tau_e = 23\text{h}, 56\text{ min}, 4.1\text{ s} = 86,141.1\text{ s}$ [13]. Therefore, the MTS (or any object on a GEO) seems to be fixed regardless of any ground observer.

Consider an Earth-centered, Earth fixed rotating reference system $E\,(x_e, y_e, z_e)$ that rotates with constant angular velocity $\boldsymbol{\omega}_e = \omega_e \hat{\boldsymbol{z}}_e$ ($\omega_e = \frac{2\pi}{\tau_e}$) with respect to an Earth-centered inertial frame system $I\,(x_I, y_I, z_I)$, both with common origin O at the Earth's center of mass, as shown in Fig. 1. Thus, a common GEO becomes a fixed point (GEO point) with respect to the rotating reference system $E\,(x_e, y_e, z_e)$. As shown in Fig. 1, the $x_e - y_e$ and $x_I - y_I$ planes coincide with the equatorial plane of the Earth, and the z_e

IOP Conf. Series: Journal of Physics: Conf. Series **911** (2017) 012006 doi:10.1088/1742-6596/911/1/012006

and z_I axes are aligned with the Earth's rotational axis. Furthermore, the x_I axis points in the vernal equinox direction, and the x_e axis points to the GEO point and coincides with the x_I axis at $t = 0$. Assuming the gravitational parameter, the magnitude of the angular velocity ω_e and the distance r_{gs} between the Earth's center and GEO point equal to unity, the nondimensional vector equation of motion for an ideal (i.e. perfectly reflecting) space mirror can then be written with respect to the rotating frame as [14]

$$\frac{\mathrm{d}^2\mathbf{r}}{\mathrm{d}t^2} + 2\boldsymbol{\omega}_e \times \frac{\mathrm{d}\mathbf{r}}{\mathrm{d}t} + \nabla U = \boldsymbol{a}, \tag{1}$$

where $\mathbf{r} = (x, y, z)^T$ denotes the position vector of the sunlight reflector with respect to the Earth's center of mass, as shown in Fig. 1. The two-body pseudopotential U is defined as $U = V + \varphi$, where V denotes the gravitational potential due to a perfectly spherical Earth and φ is the centrifugal potential in the rotating frame, which are given by $V = -\frac{1}{r}$, and $\varphi = -\frac{x^2+y^2}{2}$. The solar radiation pressure $\boldsymbol{a}$ in Eq. (1) is defined by $\boldsymbol{a} = \kappa\left(\hat{\boldsymbol{S}}(t) \cdot \boldsymbol{u}\right)^2 \boldsymbol{u}$, where κ is the reflector characteristic acceleration, $\boldsymbol{u}$ is the reflector normal unit vector, and $\hat{\boldsymbol{S}}(t)$ is the unit-vector in the direction of the sun-line. Eclipse seasons, i.e. shadow effects, on geostationary satellites have not be included in this work.

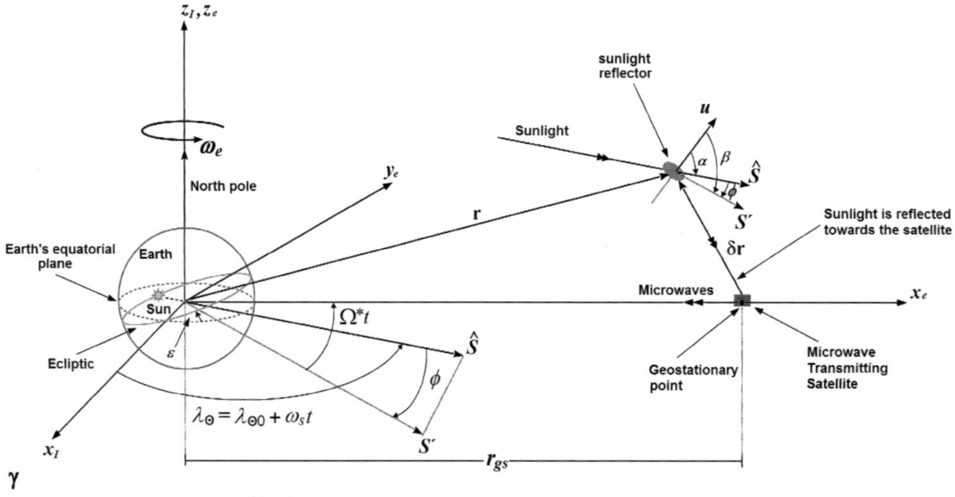

Figure 1. Rotating reference system $E(x_e, y_e, z_e)$ and inertial frame system $I(x_I, y_I, z_I)$, both with common origin O at the Earth's center of mass

Takeichi et al. [9] and Baig and McInnes [13] assumed the sun line $\hat{\boldsymbol{S}}(t)$ at a constant angle ϕ over one period. In this work, a more realistic expression for the solar perturbation $\boldsymbol{a}$ is computed, considering the path of the Sun on the ecliptic. The equatorial components of the sunline with respect to the inertial frame $I(x_I, y_I, z_I)$ are $(-\cos\lambda_\odot, -\cos\varepsilon\cos\lambda_\odot, -\sin\varepsilon\cos\lambda_\odot)^T$, where $\varepsilon = 23.5°$ is the mean Earth's ecliptic and $\lambda_\odot = \lambda_{\odot 0} + \omega_s t$ is the longitude of the Sun and $\lambda_{\odot 0}$ is the initial solar longitude as shown in Fig. 1 [15]. The sunline $\hat{\boldsymbol{S}}(t)$ in the rotating frame $E(x_e, y_e, z_e)$ can be computed as

$$\hat{\boldsymbol{S}}(t) = \begin{pmatrix} \cos t & \sin t & 0 \\ -\sin t & \cos t & 0 \\ 0 & 0 & 1 \end{pmatrix} \begin{pmatrix} -\cos\lambda_\odot \\ -\cos\varepsilon\cos\lambda_\odot \\ -\sin\varepsilon\cos\lambda_\odot \end{pmatrix} = \begin{pmatrix} \hat{S}_\xi(t) \\ \hat{S}_\eta(t) \\ \hat{S}_\zeta(t) \end{pmatrix}. \tag{2}$$

The sunlight must be redirected towards the MTS, i.e. the reflected sunlight vector must be equal to $-\delta\mathbf{r}$, as shown in Fig. 1. Therefore, the law of reflection requires that [16]

$$\boldsymbol{u} = \frac{\hat{\boldsymbol{S}}(t) + \hat{\delta\mathbf{r}}_{sr}}{\left|\hat{\boldsymbol{S}}(t) + \hat{\delta\mathbf{r}}_{sr}\right|} \tag{3}$$

Equations of motion, Eq. (1) can now be integrated numerically. To investigate bounded solutions for Eq. (1) about the GEO point, the linear ecliptic orbits found by Baig and McInnes [13] are used as initial conditions. Considering a GEO point at $\mathbf{r}_{gs} = (1, 0, 0)^T$, then the position vector $\mathbf{r}$ of the sunlight reflector can be written as $\mathbf{r} = \mathbf{r}_{gs} + \delta\mathbf{r}$, where $\delta\mathbf{r} = (\xi, \eta, \zeta)^T$ is the reflector position vector with respect to the GEO point (see Fig. 1). Assuming a constant attitude for the reflector pitch angle, a linear planar ecliptic orbit is given by [13]

$$\xi(t) = -\kappa\left(\frac{2\Omega^* + \Omega^{*2}}{\Omega^{*4} - \Omega^{*2}}\right)\cos^2\alpha\cos\beta\cos(\Omega^* t), \tag{4}$$

$$\eta(t) = \kappa\left(\frac{\Omega^{*2} + 2\Omega^* + 3}{\Omega^{*2} + 2\Omega^*}\right)\left(\frac{2\Omega^* + \Omega^{*2}}{\Omega^{*4} - \Omega^{*2}}\right)\cos^2\alpha\cos\beta\sin(\Omega^* t), \tag{5}$$

where $\zeta(t) = 0$, α is the reflector pitch angle (i.e. $\cos\alpha = \hat{\boldsymbol{S}}(t)\cdot\boldsymbol{u}$), ϕ is the inclination angle for the sun-line above the Earth's equatorial plane, $\beta = \alpha + \phi$ is the angle that the reflector normal makes with the equatorial plane of Earth, as shown in Fig. 1, and $\Omega^* = 0.9973$ is the nondimensional angular velocity of the rotating frame relative to the sun-line [13].

Figure 2 shows a numerical simulation of the variation of the sunlight reflector relative position coordinates with a characteristic acceleration $\kappa = 0.018$ mms^{-2} during one years span. The linear solutions, Eqs. (4)-(5), were used as initial conditions in the numerical simulation with $t = 0$, $\phi = 0$, $\alpha = 0$, $\phi = 0$ and $\lambda_{\odot 0} = 180°$, such that the sun-line is aligned with the vernal equinox at $t = 0$. Figures 2(a)-(c) show a bounded behavior of the reflector orbit with respect to the MTS. However, the motion is strongly affected by the solar perturbation. Figure 2(d) shows the variation of the reflector pitch angle α (i.e. control history) when the law of reflection is applied in the space mirror attitude along the time. Finally, Figs. 3(a)-(c) and Fig. 3(d) show the bounded orbits for the reflector with respect to the GEO point in the rotating frame, respectively.

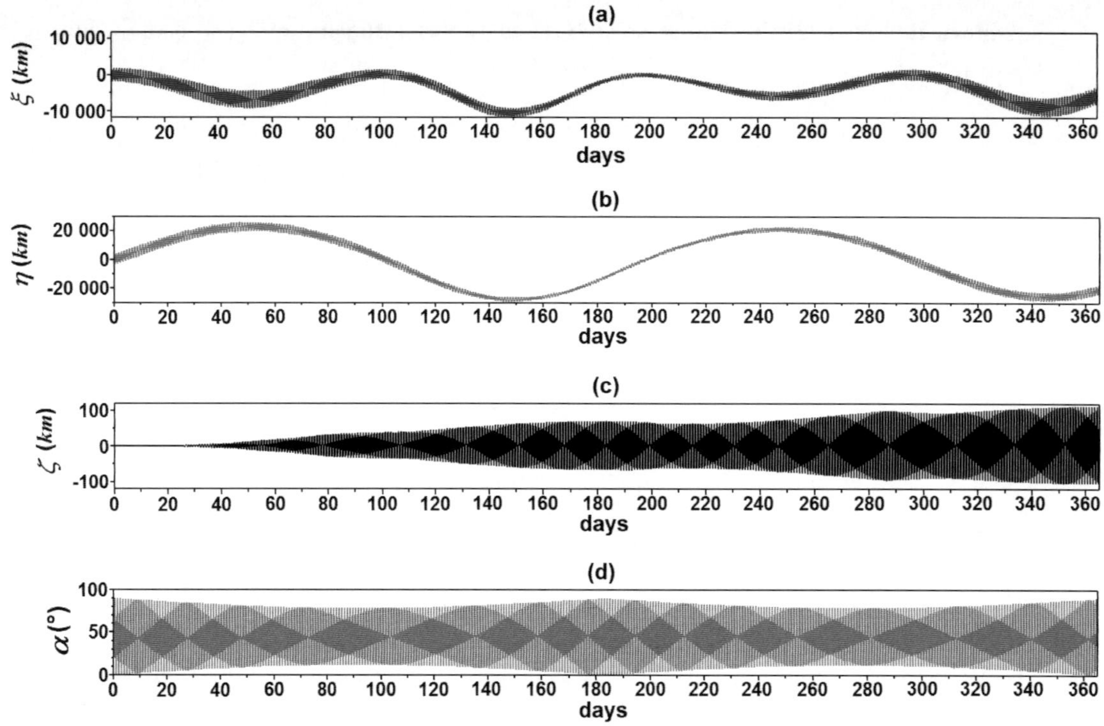

Figure 2. (a)-(c) Variation of the sunlight reflector relative position coordinates with a characteristic acceleration $\kappa = 0.018$ mms^{-2} during one year span. (d) Variation of the reflector pitch angle α (i.e. control history)

3. Conclusions

In this study a SPS system have been investigated considering a space mirror and a MTS in formation around a GEO point with respect to an Earth-centered rotating frame that rotates with the same angular velocity of the Earth. In order to include the direction of the sun-line, the position of the Sun was approximated by a circular ecliptic trajectory, assuming a mean inclination of Earth's equator with respect to the ecliptic equal to 23.5°. To redirect the sunlight to the MTS, the law of reflection was applied in the reflector attitude, so that a bounded orbit for the sunlight reflector was obtained using as initial conditions the ecliptic orbits found in a linear approximation with constant attitude for the reflector pitch angle. As was noted, solar perturbation affected the synchronization of the SPS system, increasing the longitude and producing large variations of the reflector pitch angle. The possibility of reducing the longitude of the SPS system and the variation of the reflector pitch angle using a continuous orbital control by thrusters or a tethered system will be considered in future studies.

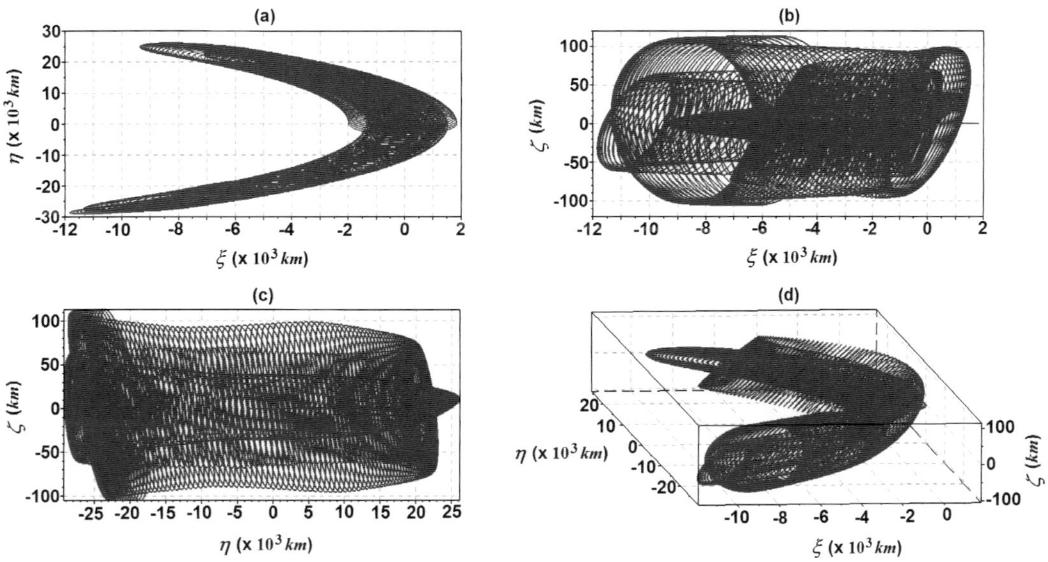

Figure 3. (a)-(c) Projections in the $\xi - \eta$, $\xi - \zeta$, and $\eta - \zeta$ planes for the sunlight reflector motion in the rotating frame corresponding to the relative position vectors in Fig. 2. (d) Bounded orbit for the sunlight reflector in the rotating frame.

Acknowledgments
Authors thank the financial support of the FAPESP (São Paulo Research Foundation, Brazil), Grants 2011/08171-3, 2013/03233-6.

References
[1] Glaser P E 1968 Power from the Sun: Its future, *Science* **162** 857-61
[2] Billman K W, Gilbreath W P and Bowen S W 1977 *Introductory Assessment of Orbiting Reflectors for Terrestrial Power Generation, NASA TM X-23* p 230
[3] Billman K W, Gilbreath W P and Bowen S W 1977 *Satellite mirror systems for providing terrestrial power: System Concept, 23rd Meeting of the AAS, Paper No. AAS-77-240* (San Francisco, CA)
[4] Ehricke K A 1979 Space light: space industrial enhancement of the solar option, *Acta Astronaut.* **6** 1515-1633
[5] Mori M, Nagayama H, Saito Y and Matsumoto H 2001 *Summary of Studies on Space Solar Power Systems of National Space Development Agency of Japan, 52nd Int. Astronaut. Congr., Paper No. IAF-01-R.1.04,* (Toulouse, France)
[6] Oda M and Mori M 2003 *Stepwise Development of SSPS, JAXA's Current Study Status of the 1GW Class Operational SSPS and Its Precursor, 54th Int. Astronaut. Congr., Paper No. IAC-03-R.3.03* (Bremen, German)
[7] Carrington C and Feingold H 2000 *SSP Systems Integration, Analysis and Modelling, SSP Tech. Interchange Meeting #3* (Huntsville, AL)
[8] Turner A E 2006 *Power Transfer for Formation Flying Spacecraft, Space 2006 Conf., Paper No. AIAA 2006-7493* (San Jose, CA)
[9] Takeichi N, Ueno H, and Oda M 2005 Feasibility Study of a Solar Power Satellite System Configured by Formation Flying, *Acta Astronaut.* **57** 698-706

[10] Fischer H and Haerting A 1992 Why Light-Levitation Geostationary Cylindrical Orbits Are Not Feasible, *J. Astronaut. Sci.* **40** 329-33

[11] van de Kolk C 1999 *Stability of Levitated Cylindrical Orbits by Using Solar Sails, AAS/AIAA Astrodynamics Specialist Conf., American Astronautical Society Paper No. 99-335* (Gridwood, AK)

[12] Wie B. and Roithmayr C M 2001 *Integrated orbit, attitude, and structural control systems design for space solar power satellites, NASA TM-2001-210854*

[13] Baig S and McInnes C R 2010 Light-Levitated Geostationary Cylindrical Orbits Are Feasible, *J. Guid. Control Dyn.* **33** 782-93

[14] McInnes C R 1999 *Solar Sailing: Technology, Dynamics And Mission Applications* (London: Springer Praxis) pp 250-56

[15] Krivov A V and Sokolov L L 1995 Dikarev, V.V., Dynamics of Mars-orbiting dust: Effects of light pressure and planetary oblateness, *Celest. Mech. Dyn. Astron.* **63** 313-39

[16] Salazar F J T, McInnes C R and Winter O C 2016 Periodic orbits for space-based reflectors in the circular restricted three-body problem, *Celest. Mech. Dyn. Astron.* doi:10.1007/s10569-016-9739-3

IOP Conf. Series: Journal of Physics: Conf. Series **911** (2017) 012007 doi:10.1088/1742-6596/911/1/012007

Studying the energy variation in the powered Swing-By in the Sun-Mercury system

A F S Ferreira[1], A F B A Prado[1], O C Winter[2] and D P S Santos[3]

[1]Instituto Nacional de Pesquisas Espaciais, INPE, São José dos Campos, Brazil
[2] Faculdade de Engenharia - Universidade Estadual Paulista, FEG-UNESP, Guaratinguetá, Brazil
[3] Universidade Estadual Paulista, UNESP, São João da Boa Vista, Brazil

E-mail: aleferrazsilva@hotmail.com

Abstract. A maneuver where a spacecraft passes close to Mercury and uses the gravity of this body combined with an impulse applied at the periapsis, with different magnitudes and directions, is presented. The main objective of this maneuver is the fuel economy in space missions. Using this maneuver, it is possible to insert the spacecraft into an orbit captured around the Sun or Mercury. Trajectories escaping the Solar System are also obtained and mapped. Maps of the spacecraft energy variation relative to the Sun and the types of orbits resulting from the maneuver are presented, based in numerical integrations. The results show that applying the impulse out of the direction of motion can optimize the maneuver due to the effect of the combination of the impulse and the gravity.

1. Introduction

Mercury is the smallest and most internal planet in the Solar System. It has an equatorial radius of approximately 2440 km, 0.055 times the Earth's mass and an orbital period around 88 days. It is approximately 57,910,000 km away from the Sun and, among all the planets of the solar system, it is the one with the largest eccentricity, in the order of 0.2056 [1].

Missions such as Mariner 10 and Messenger were sent for observations of Mercury [2-5]. Along the way, Swing-By maneuvers were made to help the approach of the probes to the destination and then to insert them into the planet's orbit. Mariner 10 was the first probe to use the gravity assisted maneuver with a celestial body in a mission. Both probes had close encounters with the planet for observations and data collection.

Swing-By maneuvers or gravity assisted maneuvers are techniques that use a close approach of a spacecraft with a celestial body to give or to remove energy of the spacecraft. The powered Swing-By maneuver combines the gravity of the celestial body with the application of an impulse during the close encounter to change the trajectory of the spacecraft, and it is recommended in situations where only the gravity part the maneuver is not sufficient to meet the mission needs.

There are works about this subject available in the literature. Prado (1996) [6] studied the powered Swing-By maneuver and its efficiency for systems in circular orbits. In 1999, Casalino et al. [7] made an analytical extension of Prado's work, considering the impulse applied in different positions of the orbit. The detailed result for the case with the impulse applied in the periapsis of the orbit of the spacecraft, including an analysis of the maximum gain and loss of energy, was presented in [8].

Ferreira et al. (2017a) [9] made a study detailed showing the effects of the eccentricity of the orbit of the primary bodies of the system in the maneuver. A numerical map of the energy gain due to the powered Swing-By maneuver, for the circular case and assuming different magnitudes, directions and positions of the impulse, was present in [10].

The present paper is an application of this maneuver, for the Sol-Mercury system, based in the study made in references [9] and [10]. The Sun-Mercury system is chosen because it has high eccentricity and the fact that it is the focus of many studies and missions. The main reason to perform the present study, and not only using the results shown in references [9-10], is that important aspects, like the occurrence of collisions and escapes, are highly dependent on the particular system. The exact values of the mass and dimensions of the primaries has strong effects in the behavior of the trajectories, and general results like the ones shown in references [9-10] give only a first view of the problem. Therefore, the present publication looks in detail the Sun-Mercury case, skipping general description and studies already published in the cited papers.

2. Statement of the problem

The work was developed by numerically integrating the equations of motion of the spacecraft, based in the Elliptical Restricted Three-Body Problem (ERTBP) [11]. The advantage of this model is the increase of the accuracy in the results, since it does not depend on approximations made by the "patched-conics" or circular models. It also helps to identify trajectories resulting in captures and collisions. We consider M_1 the Sun, M_2 Mercury and M_3 the spacecraft with negligible mass and that moves under the gravitational forces of the two primaries. The equations of motion [11] are given by $\ddot{x} = \frac{-(1-\mu)(x-x_1)}{r_1^3} - \frac{\mu(x-x_2)}{r_2^3}$ and $\ddot{y} = \frac{-(1-\mu)(y-y_1)}{r_1^3} - \frac{\mu(y-y_2)}{r_2^3}$, where x_1, y_1 are the Sun positions coordinates, x_2, y_2 represents the Mercury position coordinates; μ is the mass parameter of M_2; $r_1$is the distance between the spacecraft and the Sun and r_2 is the distance between the spacecraft and Mercury.

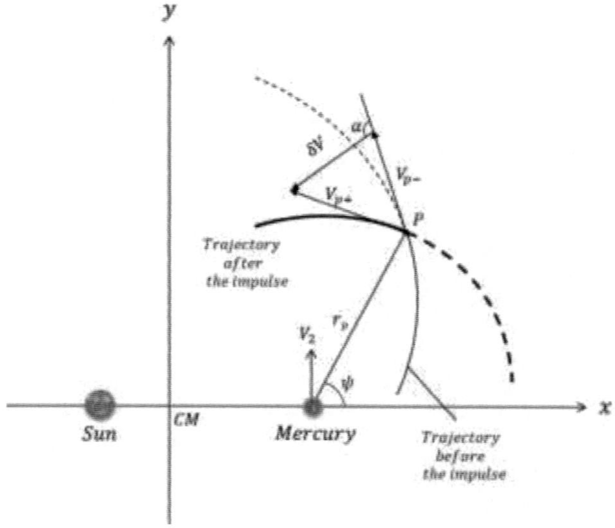

Figure 1 – Geometry of the powered Swing-By maneuver.

Figure 1 shows the geometry of the powered Swing-By maneuver. It starts from the periapsis of the trajectory of the spacecraft around Mercury, defined by the initial conditions: ψ, the approach angle; V_{inf-}, the approach speed given by $V_{inf-} = \sqrt{V_{p-}{}^2 - 2\mu/r_p}$; and r_p, the periapsis radius. The equations were numerically integrated reversely in time [12-15], without the impulse, until it reaches a distance where the effect of M_2 on the spacecraft can be neglected and the movement between the spacecraft and the Sun is considered Keplerian. Then, the two-body energy is calculated. Returning the spacecraft to the periapsis of the orbit, the impulse is applied according to the magnitude (δV) and direction (α) defined, and the trajectory is propagated forward in time until it reaches the distance where the spacecraft-Sun movement is considered Keplerian. The new orbit is then obtained. Based on the parameters before and after the maneuver, the variations generated by the maneuver can be obtained by subtracting directly the quantities involved, therefore mapping the variations of the desired quantities as a function of all variables. In this work the energy variation and the type of the orbit of the spacecraft is analyzed, considering Mercury in the periapsis and in the apoapsis of its orbit around the Sun.

3. Results

The eccentricity of Mercury has been considered and its main effect is that the velocity of Mercury (V_2) around the Sun is not constant. It varies according to its position, being faster in the periapsis and slower in the apoapsis of its orbit. The energy variation measured with respect to the main body of the system is analyzed. For positive variation values, the second orbit after the impulse increased its energy compared to the first orbit. If the variation is negative, the spacecraft loses energy in the second orbit compared to the same first orbit.

The results are obtained for the initial conditions: $\mu = 1.65 \times 10^{-7}$, $r_p = 1.1$ Mercury's radius, $V_{inf-} = 1.0$ canonical units,$\psi = 90°$ and $\psi = 270°$. Figure 2 shows the energy variations, considering Mercury positioned in the periapsis of its orbit at the time of the Swing-By. Figure 2(a) considers $\psi = 270°$, and, in this case, the gravity works for the energy gain. Note that the maximum energy variations occur for α between -45° and 0°. If the goal is to gain energy, it is more consistent to send the spacecraft towards the secondary body, to enhance the effect of gravity. Figure 2(b), made for $\psi = 90°$, has the gravity working for the energy loss. The maximum energy variations have α positive, to move the spacecraft away from the body and to minimize the effect of gravity. In general, the magnitude of the energy variation increases with the magnitude of the impulse. For impulses with1.5 c.u. and 2.0 c.u., there were cases of captures or collisions of the spacecraft. This type of analyses needs to be made for each system of primaries involved, because the results are highly dependent on the particular physical characteristics of the system under study. The minimum energy variation occurs for retrograde impulses.

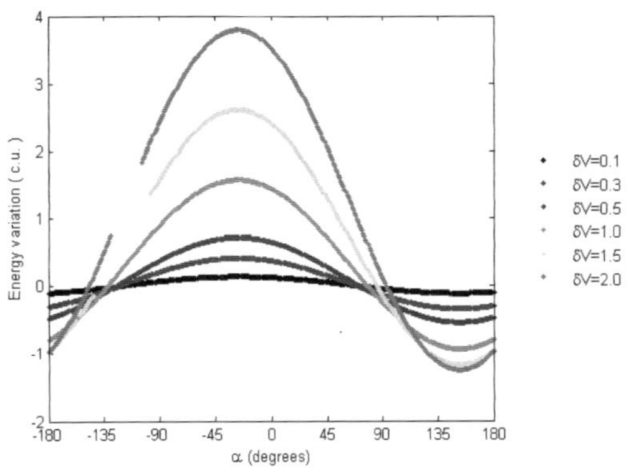

a) $\psi = 270°$

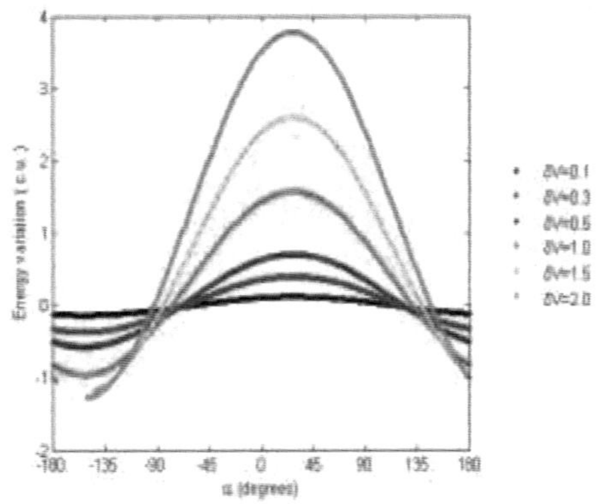

b) $\psi = 90°$

Figure 2 – Energy variation for $\nu = 0°$.

Figure 3 shows the solutions when Mercury is positioned in the apoapsis of its orbit at the time of the Swing-By. The main difference compared to Fig. 2 is the magnitude of the energy variation. Note that, even M_2 being in a position which does not favor energy gains, the combination impulse-gravity assist are more significant, providing greater energy variations. The direction to apply the impulse, in this case, is further from the direction of the spacecraft's motion.

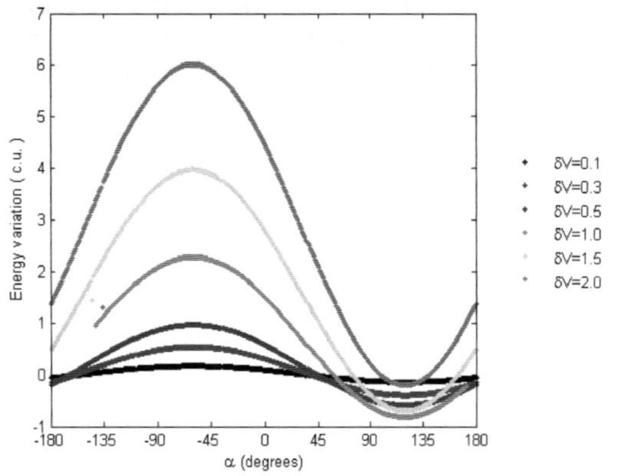

a) $\psi = 270°$

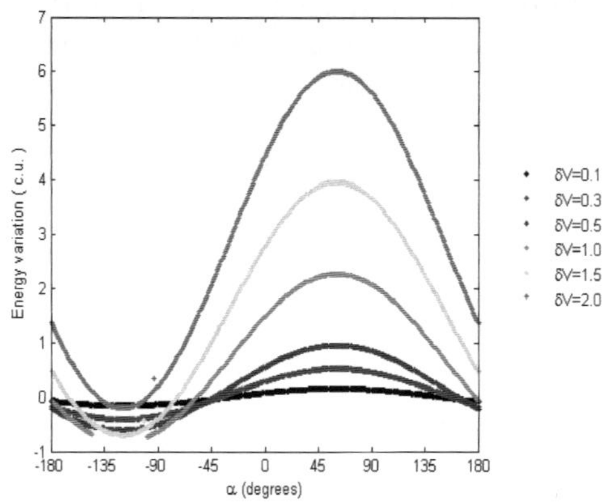

b) $\psi = 90°$

Figure 3 – Energy variation for $\nu = 180°$.

Note that the results presented in Figures 2 and 3 confirm the study presented in [9], made for a system similar to the Earth-Moon, but they add the particularities of the specific system, like their sizes and masses. Similar behaviors are noted in the energy variations. The difference in the magnitude of the energy variation and in the amount of captures and collisions are due to different size, mass and distance of the systems, as already explained.

The effect of the maneuver in the type of the orbit of the spacecraft is also presented. The descriptions are obtained from the analysis of energy (ellipse or hyperbola) and angular momentum (direct or retrograde) of the spacecraft with respect to the main body of the system.

Table 1: Orbits descriptions

After: / Before:	Direct ellipse	Retrograde ellipse	Direct hyperbole	Retrograde hyperbole
Direct ellipse	A	E	I	M
Retrograde ellipse	B	F	J	N
Direct hyperbole	C	G	K	O
Retrograde hyperbole	D	H	L	P

Each notation has a color to facilitate the visualization of the figures (Table 1). The notation "Y" (red) represents orbits where the spacecraft collide with Mercury and "Z" (black) are trajectories where the spacecraft remained orbiting Mercury until the final integration time. The orbits of types "A", "B", "E" and "F" are closed orbits that remain closed after the impulse. They may or may not have their directions of movement modified. The orbits of type "C", "D", "G" and "H" are orbits where the spacecraft was captured by the main body after the impulse. They start with an open orbit (hyperbole) that becomes an ellipse. Since types "I", "J", "M" and "N" represent orbits where the spacecraft escaped from the main body. Initially, it was in an elliptical orbit and, after the maneuver, it went into a hyperbolic orbit. Cases "I" and "N" maintained the direction of movement. Finally, the orbits of type "K", "L", "O" and "P" are open orbits that remained open, in a hyperbole, after the maneuver. The results show maps of the orbits having impulses ranging from 0 to 2 c.u. on the horizontal axis and, in the vertical axis, the angle defining the direction of the impulse, ranging from -180° to 180°.

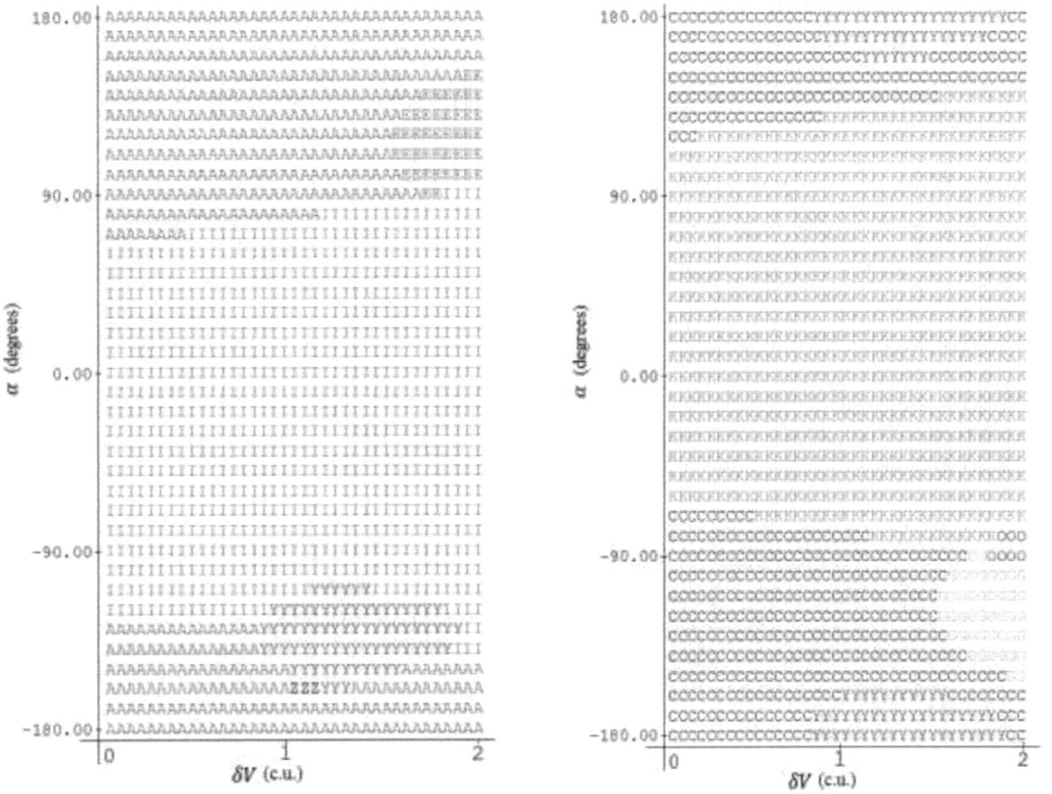

a) $\psi = 270°$ b) $\psi = 90°$

Figure 4 – Type of the orbits for $\nu = 0°$.

Figure 4(a) shows cases where the spacecraft escaped from the main body after the maneuver ("I"), being α approximately between -90° and 90°. In these cases the impulses are prograde. This behavior is consistent considering that $\psi = 270°$ is the region of energy gains due to gravity part of the maneuver. The gaining of more energy results in larger velocities, therefore the greater the chance to generate an escape for the spacecraft. The few cases where the spacecraft collided with the body ("Y") are for $\delta V > 1.0$ c.u. and with an impulse with a component opposite to the movement of the spacecraft and the other component in the direction of the secondary body. There are also orbits of type "A", which are direct ellipses, and type "E", which are ellipses that change the direction of movement after the maneuver; and finally trajectories of type "Z", which makes the spacecraft to be captured by Mercury.

For the case $\psi = 90°$, the gravity works to reduce energy, increasing the chances of capture. Figure 4(b) shows trajectories where the spacecraft was captured by main body after the maneuver ("C" and "G"). Note that the orbits resulting in capture have retrograde impulses ($\alpha > 90°$ and $\alpha < -90°$), which decelerates the spacecraft and works in favor of the gravity to decrease energy. The orbits of type "G" are captures with direction of the movement inverted after the maneuver. Some cases of collisions with M_2 ("Y") also occurred, as well as orbits of types "K" and "O". In orbits of type "K", the spacecraft remained in the same type of trajectory after the maneuver. They occur for impulse direction greater than -90° to just over 90°, and for all magnitudes of the impulse.

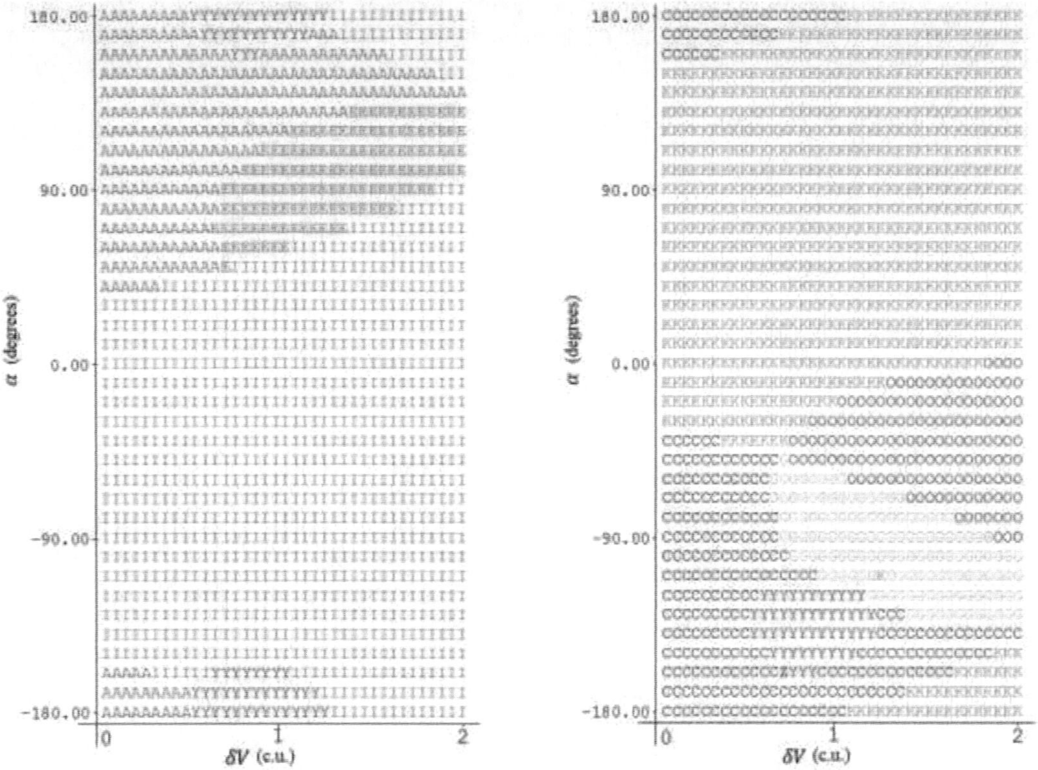

a) $\psi = 270°$ b) $\psi = 90°$

Figure 5 – Type of the orbits for $\nu = 180°$.

In Figure 5, the main change was the position of Mercury, which is now in the apoapsis of its orbit around the Sun. Figure 5(a) has the same orbit type as Figure 4(a), but the impulse configuration for these orbits has some differences. The region with direct ellipse ("A") is smaller, while the region with direct ellipse that has become retrograde ("E") has increased. There were some cases of spacecraft collisions with Mercury ("Y") and all others were escapes from the main body ("I"). Figure 5(b) shows trajectories where the spacecraft was captured by the main body after the maneuver ("C" and "G"). Some cases of collisions with Mercury and orbits of types "K" and "O" also occurred. In orbits of the type "K", the spacecraft remained in the same type of trajectory after the maneuver and in the type "O" only the direction of movement was inverted.

Associating the energy variations with the types of orbits, we see that when $\psi = 270°$, for both positions of Mercury in orbit, the trajectories with maximum energy variation, the spacecraft escaped from the Sun after the complete maneuver. When $\psi = 90°$, in the trajectories with maximum energy variation, the spacecraft came from an open orbit with respect to the Sun and remained in an open orbit with the same direction of the first one after the maneuver.

4. Conclusions

A study of the behavior of the spacecraft energy in a trajectory that performs a powered Swing-By maneuver was presented. Different initial conditions were simulated, considering Mercury in the periapsis and apoapsis of its orbit around the Sun and with the orientation of the Swing-By in the region of maximum gains and losses of energy due to the gravitational part of the maneuver.

The results show the energy variation and the characteristic of the impulse for each solution. For some cases there are captures or collisions of the spacecraft by Mercury. The types of trajectories performed by the spacecraft before and after the maneuver, with respect to the main body, were also presented. For the cases simulated here, there are trajectories where the spacecraft escaped from the main body and others where it was captured. Collisions with Mercury were also detected and they represent a danger for the mission. This type of risk can only be mapped by studying the specific system of primaries, since general results can only give an idea of the gains and losses of energy.

In general, the solutions show that applying the impulse out of the direction of motion can optimize the maneuver due to the effect of the combination of the impulse and the gravity.

Acknowledgements

The authors wish to express their appreciation for the support provided by the National Council for the Improvement of Higher Education (CAPES), the National Institute for Space Research (INPE), the grants # 406841/2016-0 and 301338/2016-7 from the National Council for Scientific and Technological Development (CNPq), and the grants #2011/08171-3, 2016/14665-2, from Sao Paulo Research Foundation (FAPESP).

References

[1] NASA, JPL. Planets – Mercury: by the numbers. Site manager: Alice Wessen. Editor: Phillips Davis. Science Writer: Bill Dunford. Available in: http://solarsystem.nasa.gov/planets/mercury/facts.
[2] Dunne J A, Burgess E 1978 The voyage of Mariner 10 National Aeronautics and Space Administration SP 424
[3] McNutt Jr R L, Solomon S C, Grard R, Novara M, Mukai T 2004 An international program for mercury exploration: synergy of Messenger and Bepicolombo *Adv. Space Res.* **33** pp 2126–2132
[4] Mcnutt Jr R L, Solomon S C, Gold R E, Leary J C 2006 The messenger mission to mercury: development history and early mission status *Adv. Space Res.* **38** pp 564–571
[5] NASA, Messenger Trajectory, Developed by JHU/APL, 1999-2016, Available in: http://messenger.jhuapl.edu/the mission.
[6] Prado A F B A 1996 Powered Swing-By *J. Guid. Control Dynam.* **19** 5 pp 1142- 1147
[7] Casalino L, Colasurdo G, Pastrone D 1999a Simple Strategy for Powered Swing-By *J. Guid. Control Dynam.* **22** 1 pp 156-159
[8] Ferreira A F S, Prado A F B A, Winter, O C 2015 A numerical study of powered Swing-Bys around the Moon *Adv. Space Res.* **56** 2 pp 252-272
[9] Ferreira A F S, Prado A F B A, Winter O C, Santos D P S 2017a Effects of the eccentricity of the primaries in a powered Swing-By maneuver *Adv. Space Res.* **59** 8 pp 2071-2087 Doi: 10.1016/j.asr.2017.01.033
[10] Ferreira A F S, Prado A F B A, Winter O C 2017b A numerical mapping of energy gains in a powered Swing-By maneuver *Nonlinear Dynam.* **89** pp 791-818 Doi: 10.1007/s11071-017-3485-2.
[11] Szebehely V 1967 *Theory of Orbits* (New York: Academic Press)
[12] Vieira Neto E, Winter O C 2001 Time analysis for temporary gravitational capture: satellites of Uranus *Astron J.* **122** 1 pp 440-448
[13] Prado A F B A 2002 Numerical Study and Analytic Estimation of the Forces Acting in the Gravitatonal Capture *J. Guid. Control Dynam.* **25** 2 pp 368-375
[14] Neto E V, Prado A F B A 1998 Time-of-flight analyses for the gravitational capture maneuver *J. Guid. Control Dynam.* **21** 1 pp 122-126
[15] Machuy A L, Prado A F B A, Stuchi T J 2007 Numerical study of the time required for the gravitational capture in the bi-circular four-body problem *Adv. Space Res.* **40** 1 pp 118-124

IOP Conf. Series: Journal of Physics: Conf. Series **911** (2017) 012008 doi:10.1088/1742-6596/911/1/012008

Searching for orbits around the triple asteroid 2001SN$_{263}$

M P O Cavalca[1], A F B A Prado[1], J K S Formiga[2] and V M Gomes[3]

[1]National Institute for Space Research - INPE, São José dos Campos (SP), Brazil
[2]São Paulo State University - UNESP/ICT, São José dos Campos (SP), Brazil
[3]São Paulo State University - UNESP/FEG, Guaratinguetá (SP), Brazil

E-mail: mpocavalca@hotmail.com

Abstract. The asteroid 2001SN$_{263}$ is one of the possible targets of a proposed mission that would be the first Brazilian exploration in deep space, the Aster Mission. This asteroid is composed by three bodies: Alpha, Beta, and Gamma, in decreasing order of mass. For this study, it is proposed to split this triple system in two double systems: Alpha-Beta-spacecraft and Alpha-Gamma-spacecraft, all of them considered to be points of mass, such that it is possible to use the circular planar restricted three-body problem as the mathematical model. The goal is to find orbits that can be used by a spacecraft to observe the bodies Beta and Gamma. Each orbit can be identified by the initial conditions of the spacecraft with respect to Beta or Gamma: position and velocity. These orbits are classified by the minimum average distance spacecraft-celestial body. The results showed stable orbits around Beta and Gamma, with an average distance below 1.5 km, under the influence of the gravity of Alpha and the solar radiation pressure.

1. Introduction

In recent years, studies and missions related to asteroids have become recurrent and they account for a new space race [1]. Several missions were made or are under preparation to be launched. Several types of bodies have been considered in the literature, including single, double and even triple systems, like the one proposed by the Aster Mission [2-3], where the main target of study is the triple asteroid 2001SN$_{263}$ [4]. The triple system of asteroids 2001SN$_{263}$ is composed by three bodies, denominated by Alpha, Beta and Gamma, in mass descending order. Alpha is the main body, orbited by Gamma, which is located at a short distance from Alpha, below 4 km. Beta also orbits Alpha, but at a larger distance [5-6], above 16 km. The study of this triple system will be performed dividing it into two double systems: Alpha-Beta-spacecraft and Alpha-Gamma-spacecraft. The mathematical model used for the study of the two systems mentioned above is the restricted circular planar three-body problem [7], where it is assumed that Alpha, the most massive body, is orbited by Beta and Gamma, but one at a time, and that there is a spacecraft with negligible mass orbiting the system. It's also included in the mathematical model the solar radiation pressure [8]. This force depends on the area-to-mass ratio (A/m) of the spacecraft and the initial angular position of the Sun (ν).

This paper proposes to map orbits [9-10] that can be used by a spacecraft around Beta and Gamma. Each orbit can be identified by the initial conditions of the spacecraft with respect to Beta or Gamma: position and velocity. The orbits will be sorted by the criterion of minimizing the average distance spacecraft-celestial body, also used in [11]. The results presented here are an approximation, since the equations of the planar restricted three-body problem were used to model the problem, which does not take into account the irregular shapes of the bodies and the perturbation from one smaller body in the orbits around the other one.

2. Metodology

The mathematical model used to describe the motion of the spacecraft is the circular planar restricted three-body problem, which is presented in [7]. The central body, Alpha, is designated by M_1; the smaller body, Beta or Gamma, depending on the smaller asteroid considered, is called M_2; and the spacecraft, assumed to have a negligible mass, which is designated by M_3. In each system, M_1, M_2, and M_3 move in the same plane; the bodies M_1 and M_2 rotate around their center of mass in circular orbits, and M_3 travels around the bodies M_1 and M_2 influenced by the potential F_G, generated by them. Added to this dynamical model, the solar radiation pressure is added, and it corresponds to the force F_{SRP} exerted by the light of the Sun, as it directly hits the surface of a space vehicle. This force depends on the relationship between the cross-sectional area of the spacecraft and its mass (A/m); of the surface reflectivity coefficient (C_R); of the solar flux (P_S); and the average distance between the system under consideration and the Sun (r_S). When the solar radiation pressure is considered, the value of A/m is assumed to be 0.01 m^2/kg, a value compatible with the Aster mission, where it is proposed a spacecraft with mass 150 kg [3] and an area near 1.5 m^2. It is also considered the initial angular position of the Sun (ν) with respect to the system being studied. Equation 1 presents the complete dynamic model:

$$F = F_G + F_{SRP} = G\left(-\frac{m_1}{r_1^2}\hat{r}_1 - \frac{m_2}{r_2^2}\hat{r}_2 \right) - C_R P_S \frac{A}{m}\frac{1}{r_S^2}\hat{r}_S \tag{1}$$

where m_1 and m_2 are the masses of M_1 and M_2, r_1 is the distance between M_3 and M_1, and r_2 is the distance between M_3 and M_2. The criterion of the average distance minimization is to obtain the initial conditions of the spacecraft, for each orbit around M_2, that gives the lowest average distance between these two bodies. The mean distance between M_3 and M_2 for a given time interval (T) is shown in Equation 2.

$$r_{avg} = \frac{1}{T}\int_0^T r_2(t)dt \tag{2}$$

Where r_2 is the distance between M_3 and M_2. Then, it is possible to classify the orbits in order to obtain the initial conditions that minimize the distance between the two bodies and this is shown by the color maps presented in Figures 2 and 4. They show the distribution of the mean distances as a function the initial distance (D) from the vehicle with respect to the moon and the angle of reference (θ), for fixed components of the velocity v_x and v_y.

Next, the algorithm to find the orbits is described. Initially, it is assumed that Alpha (M_1) and Beta or Gamma (M_2) are aligned in an inertial frame and the spacecraft (M_3) is within a certain distance from the moon. Figure 1 shows the position of M_1, M_2, and M_3 with respect to an inertial frame and the initial conditions of M_3 with respect to M_2, which identifies each orbit. The initial angular position of the Sun is given by ν. This angle defines the direction of the solar radiation pressure: zero is when the Sun is aligned to the right and on the same line as M_1 and M_2. Other values are obtained considering the counterclockwise direction for increasing ν. Using Figure 1, it is also possible to determine the initial position and velocity of the spacecraft and then numerically integrate the equations of motion for the time interval of 30 days. The numerical method used to integrate these equations is the Runge-Kutta 78 and the integration step is variable. After each step, it is calculated if there was a collision between M_3 and M_2, and then the average distance between them is calculated. The next step is to classify the orbits according to the average distance to identify the initial conditions that keep the spacecraft in the smallest average distance from the moon.

3. Results and Discussion

The method presented here search for the initial conditions for orbits that survive for up to 30 days and that can be used by a spacecraft to travel around Beta and Gamma. It generates a large number of results. The results that will be presented here are those that are more interesting with respect to the lowest values of average distance between the spacecraft and Beta or Gamma and also those results related to the interference of the solar radiation pressure in the orbits. All the initial conditions and parameters were chosen by these criteria. Table 1 shows the numerical data of the bodies Alpha, Beta, and Gamma: mass, average diameter, semi-major axis, and orbital period used in the simulations. Initially, the orbits around Beta will be studied, and then this study is extended to Gamma.

Table 1. Physical and orbital components of the three bodies of the Asteroid 2001SN$_{263}$. [4]

Celestial body	Average diameter (km)	Mass (10^{10} kg)	Semi-major axis	Orbital period
Alpha	2.60	917.466	1.99 UA*	2.80 yr*
Beta	0.780	24.039	16.663 UA**	6.225 d**
Gamma	0.580	9.773	3.804 UA**	0.686 d**

*Around the Sun
**Around Alpha

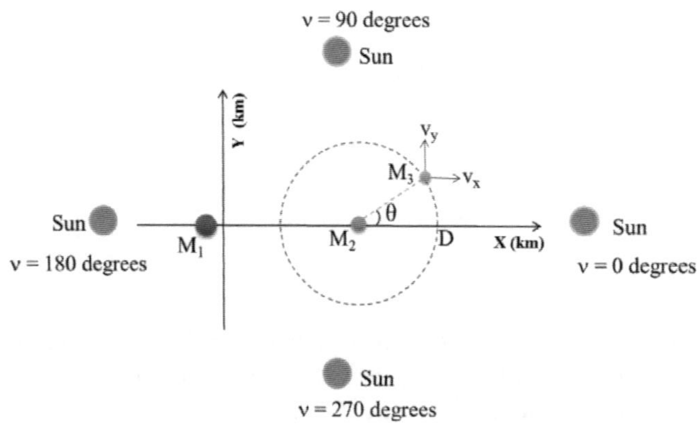

Figure 1. Geometry of the problem involving the three bodies and the initial conditions.

3.1. Study of orbits around Beta

Figures 2(a) to 2(c) show the color maps for orbits around Beta. They show the distribution of mean distances with the velocity components fixed v_x = -0.1 m/s and v_y = -0.07 m/s. The angle θ varies from 292 to 320 degrees and the initial distances vary from 0.874 to 1.20 km. In Figure 2(a) the solar radiation pressure is not considered, and in Figures 2(b) and 2(c) the solar radiation pressure is considered with A/m = 0.01 m^2/kg for ν equal to 180 and 270 degrees, respectively. As already explained, those values are used based in previous simulations, which showed that those values give orbits closer to the smaller asteroid considered. The initial conditions indicating collisions between the spacecraft and Beta in less than 30 days are indicated by the blank regions in the figures. Figures 2(a) to 2(c) shows that the largest average distances occurs for the highest values of D, the initial distance, and the smallest mean distances occur for the lowest values of D. The solar radiation pressure substantially reduces the sets of initials conditions that generate orbits around Beta. This is proven by comparing the size of the white regions. In Figure 2(a), which does not consider the radiation pressure, this region corresponds to 1/4 of the total area of the graph, whereas in Figures 2(b) and 2(c), which consider the radiation pressure, this region covers about 2/3 of the total area. The figures have their minimum and maximum mean distance values: 0.874 and 1.476 km, 1.197 and 1.480 km, 0.968 and 1.430 km, respectively.

Next, a set of initial conditions of each color map is chosen. Figures 2(a) to 2(b) draw the trajectories around Beta. Figures 3(a), 3(b), and 3(c) show these trajectories, which have the velocities equal to v_x = -0.1 m/s and v_y = -0.07 m/s and the respective values of D and θ equal to: 0.97 km and 303 degrees, 1.12 km and 306 degrees, 1.01 km and 294 degrees. Beta is fixed at the origin of the reference system, and it is represented by the circular form. The circular shape for Beta was adopted for being the simplest and also consistent with the model.

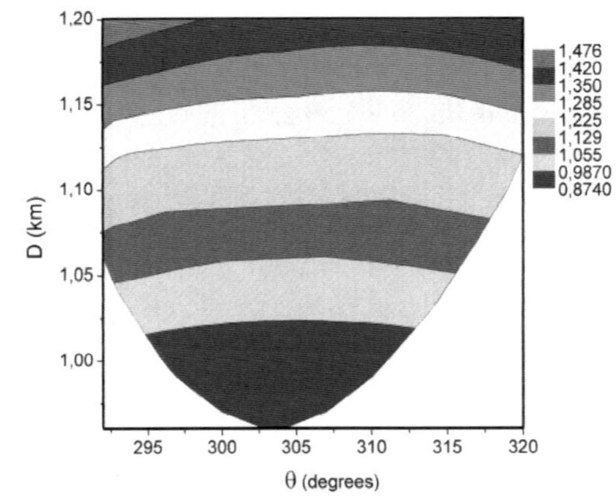

(a) A/m = 0 (without srp)

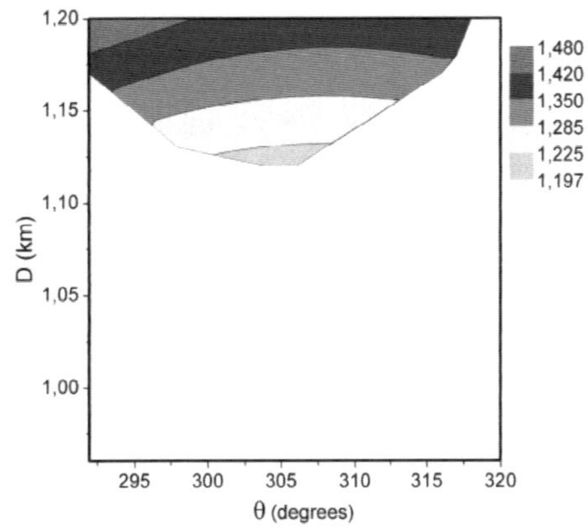

(b) A/m = 0.01 m^2/kg and ν = 180 degrees

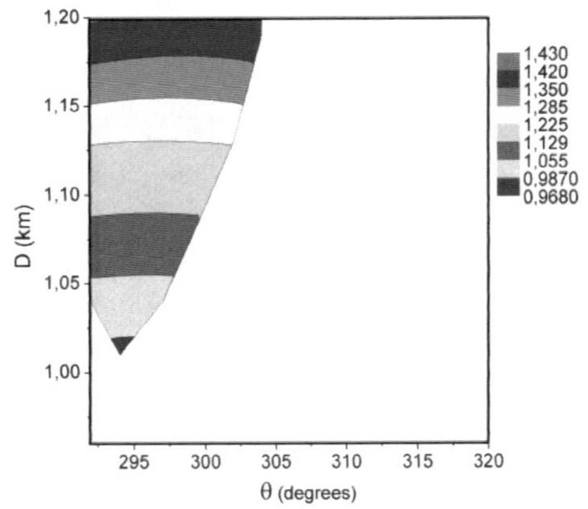

(c) A/m = 0.01 m^2/kg and ν = 270 degrees

Figure 2. Distribution of the average distance values as a function of D (km) and θ (degrees) for fixed velocity values in v_x = -0.1 m/s and v_y = -0.07 m/s.

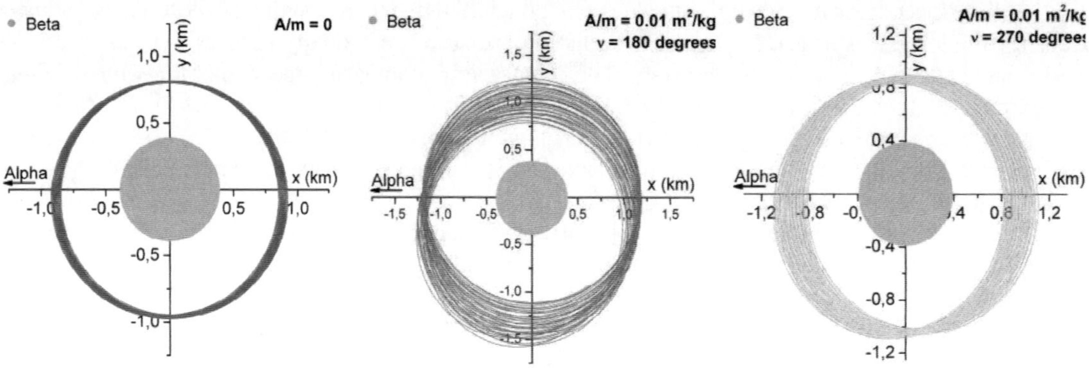

(a) r_{avg} = 0.891231 km
D = 0.97 km and θ = 303 degrees

(b) r_{avg} = 1.197299 km
D = 1.12 km and θ = 306 degrees

(c) r_{avg} = 0.969635 km
D = 1.01 km and θ = 294 degrees

Figure 3. Trajectories around Beta for fixed velocity values in v_x = -0.1 m/s and v_y = -0.07 m/s.

Figures 3(a), 3(b) and 3(c) present three trajectories around Beta, which turn in the retrograde, clockwise direction. It turns out that they are quite stable and very close to Beta, at least 800 m away. In Figure 3(a), where only the gravitational interaction between Alpha and Beta is being considered, with A/m = 0, the average distance is 891.231 m. It is important to remember that Beta is about 38 times smaller than Alpha and it is located less than 17 km from it and, even then, these characteristics are sufficient for the existence of fairly cohesive orbits. For Figure 3(b), where the solar radiation pressure is considered with A/m = 0.01 m^2/kg and ν = 180 degrees, the average distance is 1.197299 km. In this case the orbit appears to deform transversely, in the direction of the y-axis, and no longer presents such a cohesive profile, especially in the upper and lower parts of the orbit. In Figure 3(c), where the solar radiation pressure is taken with A/m = 0.01 m^2/kg and ν = 270 degrees, the mean distance is 969.635 m, and the orbit deforms in the longitudinal direction of the x-axis. It also does not have such a cohesive profile, but now it occurs, particularly, on its sides. It behaves like the last two orbits had been "flattened" by the solar radiation pressure. It is possible to conclude, for these two orbits around Beta, that the solar radiation pressure deforms the shape of the trajectory in the transverse direction for ν = 180 degrees, and in the longitudinal direction for ν = 270 degrees, that is, the deformation of the orbit, for these two examples, is perpendicular to the initial angular position of the Sun (ν) for a period of 30 days.

3.2. Study of orbits around Gamma

Figures 4(a) to 4(c) show the color maps for orbits around Gamma. They show the distribution of mean distances where the velocity components are fixed in v_x = -0.04 m/s and v_y = -0.07 m/s. The angle θ varies from 295 to 303 degrees and the initial distances vary from 781 to 910 m. In Figure 4(a) the solar radiation pressure is not considered, and in Figures 4(b) and 4(c) the solar radiation pressure is considered with A/m = 0.01 m^2/kg for ν equal to 90 and 180 degrees, respectively. The initial conditions indicating collisions between the spacecraft and Gamma in less than 30 days are indicated by the blank regions delimited by the edges of the figures.

IOP Conf. Series: Journal of Physics: Conf. Series **911** (2017) 012008 doi:10.1088/1742-6596/911/1/012008

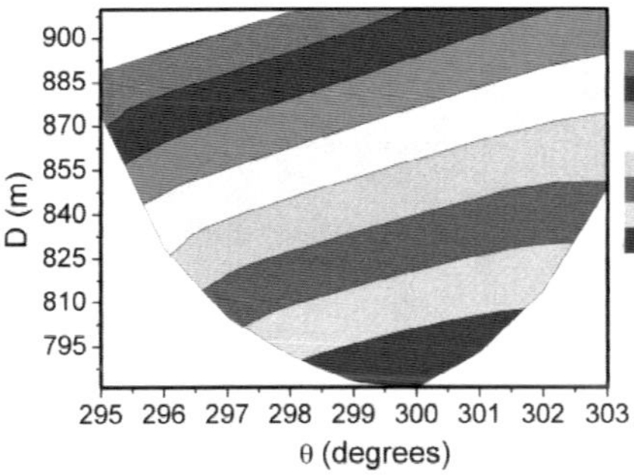

(a) A/m = 0 (without srp)

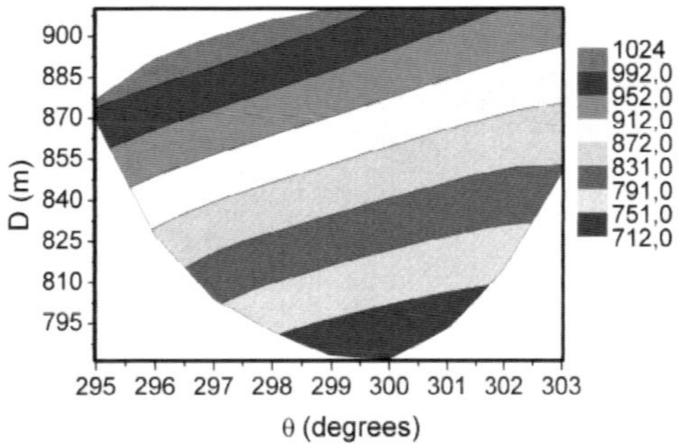

(b) A/m = 0.01 m^2/kg and ν = 90 degrees

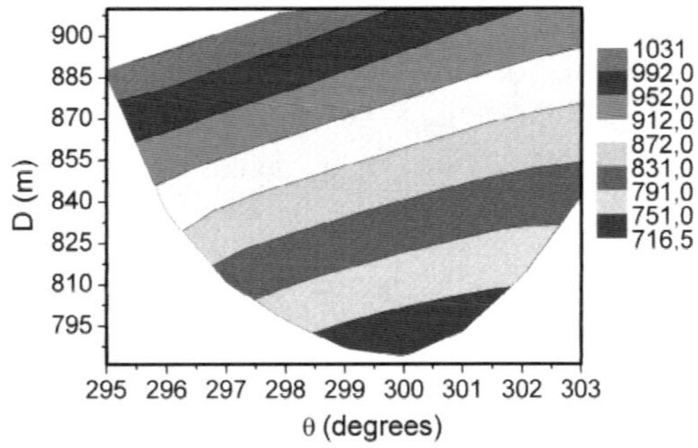

(c) A/m = 0.01 m^2/kg and ν = 180 degrees

Figure 4. Distribution of the average distance values as a function of D (m) and θ (degrees) for fixed velocity values in v_x = -0.04 m/s and v_y = -0.07 m/s.

Visually, Figures 4(a), 4(b), and 4(c) do not show large differences among them. The mean distance distribution occurs continuously varying from the smallest to the largest, as the initial distance D increases. The figures differ in their minimum and maximum mean distance values: 711 and 1033 m, 712 and 1024 m, 716 and 1031 m, respectively. The regions with the largest differences are located near the values of D between 860 and 890 m and θ between 295 and 296 degrees.

Figures 5(a), 5(b) and 5(c) show trajectories around Gamma considering the same set of initial conditions for the cases with and without radiation pressure, A/m = 0 and A/m = 0.01 m^2/kg, for ν equal to 90 e 180 degrees, respectively. The initial conditions are: D = 787 m, θ = 299 degrees, v_x = -0.04 m/s and v_y = -0.07 m/s. Gamma is fixed at the origin of the reference systems, represented by a circular form in scale. The position of Alpha is indicated by an arrow to the left on the x-axis.

Figures 5(a), 5(b) and 5(c) present three trajectories rotating in a retrograde (clockwise) direction around Gamma. The profile of the trajectories is not as cohesive as those found around Beta, but they present average distances below 735 m. Figure 5(a) considers only the gravitational interaction between Alpha and Gamma and the average distance is 733.3133 m. Gamma is 100 times smaller than Alpha and it is located less than 4 km from it, so the trajectories suffer more influence from Alpha, compared to Beta. For the values of ν of 90 and 180 degrees, respectively, the values of the mean distance were 731.8501 and 733.0861 m. The average distances did not show large differences, but it is noted that these values decreased and also a change of pattern of rotation in the format of the trajectories is observed compared to Figure 5(a), where the radiation pressure is not considered.

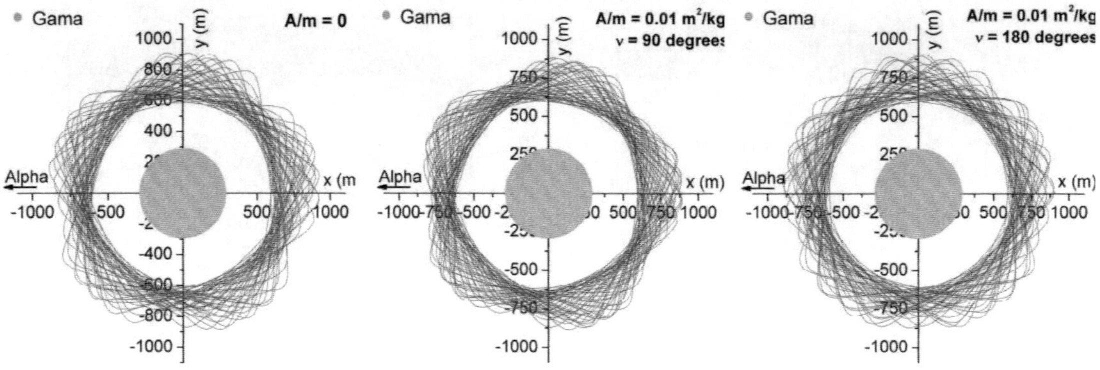

(a) r_{avg} = 733.3423 m (b) r_{avg} = 731.8501 m (c)r_{avg} = 733.0861 m

Figure 5. Trajectories around Gamma for initial conditions: D = 787 m, θ = 299 degrees, v_x = -0.04 m/s, and v_y = -0.07 m/s.

4. Conclusion

Using the algorithm developed here, based in the circular planar model of the restricted three body problem and solar radiation pressure, it was found orbits around Beta and Gamma in the respective systems Alpha-Beta and Alpha-Gamma. It was also identified initial conditions that minimize the average distance between Beta and Gama and the spacecraft using color maps. It was also possible to observe the influence of the solar radiation pressure on the shape of the trajectories and on the mean distance values for the Beta and Gamma. The results for the 30 day time period showed stable orbits around Beta and Gamma, with average distances below 1.5 km.

Acknowledgements

The authors acknowledge the financial support from CAPES; CNPq by grants 406841/2016-0 and 301338/2016-7 and FAPESP by grants # 2011/08171-3, 2016/14665-2, 2016/15675-1 and 2016/14665-2.

References

[1] ESA 2017 Missions to asteroids.http://sci.esa.int/rosetta/54342-missions-to-asteroids/. Accessed 15 March 2017

[2] Sukhanov A, Velho H, Macau E, and Winter O C 2010 The Aster Project: Flight to a near-earth asteroid *Cosmic Res.+*, pp 443–450

[3] Macau E, Winter O C, Velho H, and Sukhanov A 2012 Missão Aster: Indo aonde ninguém jamais foi *Revista Ciência Hoje*, pp 20–25

[4] Fang J, Margot J L, Brozovic M, Nolan M C, Benner L A M, and Taylor P A 2011 Orbits of near-earth asteroid triples 2001SN$_{263}$ and 1994 cc: Properties, origin, and evolution *Astron. J.*, **141** pp 154-168

[5] Masago B Y P L, Prado A F B A, Chiaradia A P M, and Gomes V M 2016 Developing the precessing inclined bi-elliptical four-body problem with radiation pressure- to search for orbits in the triple asteroid 2001SN$_{263}$*Adv. Space Res.*, **57** pp 962-982

[6] Prado A F B A 2014 Mapping orbits around the asteroid 2001SN$_{263}$ *Adv.Space Res.*, **53** pp 877–889

[7] Szebehely V 1967 *Theory of Orbits*Academic Press New York

[8] Montenbruck O and Gill E 2001 *Satellite Orbits* Springer-Verlag Berlin Heidelberg, pp 77-79

[9] Araújo R A N, Winter O C, Prado A F B A, and Sukhanov A 2012 Stability regions around the components of the triple system 2001SN$_{263}$ *Mont. Not. R. Astron. Soc.*, pp 3058–3073

[10] Araujo R A N, Winter O C, and Prado A F B A 2015 Stable retrograde orbits around the triple system 2001 SN$_{263}$ *Mont. Not. R. Astron. Soc.*, **449** pp 4404-4414

[11] Prado A F B A 2015 A dynamical study of distant retrograde orbits in moons and double asteroids of the solar system *ISTS* Kobe, Japan **1** pp 1-6

On the use of a variable coefficient of reflectivity associated with an augmented area-to-mass ratio to de-orbit CubeSats

J. B. Silva Neto[1] and A. F. B. A. Prado[1] and D. M. Sanchez[1] and J. K. S. Formiga[2]
[1]National Institute for Space Research, São José dos Campos, SP, Brazil
[2]São Paulo State University - UNESP/ICT, São José dos Campos, SP, Brazil
E-mail: jose.batista@inpe.br

Abstract. The number of missions involving CubeSats increases each year, including Brazilian missions. The simplification of the projects, aiming smaller structures and less weight, makes the missions of CubeSats cheaper compared to large satellites. Another point is that, in recent years, the concern about the increase of objects in orbit becomes important, since operational satellites and upper stage of launchers that populate the Low Earth Orbit region (LEO region) are more numerous. To try to reduce the impact of this problem, there is a consensus that objects in LEO region should re-enter in the atmosphere within 25 years. This is becoming a limit for projects of CubeSats. In this scenario, the present paper presents an alternative to this problem, which is the use of devices of area augmentation and/or variation of the coefficient of reflectivity in a CubeSat. By using these devices, it is possible, even in regions where the air density is not sufficiently large or where there is no more atmosphere, to force the decay of a CubeSat. Then, one can carry out missions of CubeSat to aim for higher altitudes than current missions, expanding the possibilities of applications.

1. Introduction

In recent years the number of missions with CubeSats has increased. The low cost of assembly and launch of CubeSats, due to the reduced weight and size, make them ideal for many different applications. For this reason CubeSats are becoming more popular in the Brazilian space program [1]. Some of these Brazilian CubeSats are already in space and others are under development. Examples of Brazilian CubeSats are ITASAT, CONASAT and NANOSATC-BR1. Along with the interest in CubeSats, not only in Brazil, but worldwide, there is a growing concern about the accumulation of objects in orbit. Among these objects are operational satellites, disabled satellites (not controlled), upper stages of launchers and fragments of collisions. To try to reduce this problem, some international organizations, such as the Inter-Agency Space Debris Coordination Committee (IADC), NASA and ESA, have issued documentation on the subject in order to regulate disposal practices for satellites at the end-of-life. Within the discussion on the subject there is a consensus that all the objects in Low Earth Orbit region (LEO region) should re-enter in the atmosphere in up to 25 years [2]. Another important concept is the de-orbit, which refers to the forced removal of any object in orbit [3]. Some of the de-orbit techniques use space tethers, thrusters, solar sails and low thrust devices. New concepts of de-orbit has been studied in recent years, one of them related to techniques that use solar radiation pressure and atmospheric drag to force the decay of a body in orbit around the Earth [4, 5, 6, 7]. These devices are less complex than a solar sail. They have variable area (like an inflatable balloon) and variable coefficient of reflectivity that, depending on its design, eliminates the need of attitude controllers. The possible elimination of the need of attitude controllers makes these type of devices ideal for use in CubeSats. At this time there are companies developing inflatable balloon devices[1]. Such devices would be inactive during the years of the mission and, at the end of the mission, a balloon is inflated in the bodies within the LEO region. The increase of the atmospheric drag force caused by the increase of the area-to-mass ratio causes the decay of the body. However, the low density of the air in high altitude limits the use of this type of device to low altitudes, or it is necessary the use of a large area increase [8]. The solution presented in this study relates the use of devices able to change its area, as the balloon mentioned above, in addition to devices with the ability to change the coefficient of reflectivity. A simple theoretical design of this device would be a balloon coated by a fabric able to change its color [9]. This type of device has a valid application in CubeSats missions and its use can increase the chances of future missions in the LEO region at higher altitudes, which could decrease the number of CubeSats in the formation of a constellation. The present paper does not address the constructive aspects of the device, but only the orbital dynamics of a CubeSat is studied. The study is done numerically, via integration of the equations of motion of a perturbed satellite. The next section shows the method used in detail.

2. Method

The study is done numerically through the integration of the equations of motion of a perturbed satellite. These equations are:

$$\begin{aligned}\ddot{\mathbf{r}} &= -\frac{GM_E}{r^3}\mathbf{r} - GM_\odot\left(\frac{\mathbf{r}-\mathbf{r}_\odot}{|\mathbf{r}-\mathbf{r}_\odot|^3} - \frac{\mathbf{r}_\odot}{|\mathbf{r}_\odot|^3}\right) - \\ & GM_M\left(\frac{\mathbf{r}-\mathbf{r}_M}{|\mathbf{r}-\mathbf{r}_M|^3} - \frac{\mathbf{r}_M}{|\mathbf{r}_M|^3}\right) + \mathbf{P}_G + \mathbf{P}_D + \mathbf{P}_{SRP} \qquad (1)\end{aligned}$$

where G is the gravitational constant, M_E, $M_\odot$, M_M are the masses of the Earth, the Sun, and the Moon, respectively. $\mathbf{r}$, $\mathbf{r}_\odot$, $\mathbf{r}_M$ are, in order, the position vector of the satellite, the Sun, and the Moon. The second and third terms of Equation 1 can be found in Reference [10] and are due to the perturbations of the Sun and the Moon, respectively. $\mathbf{P}_G$ is the acceleration due to the Geopotential, $\mathbf{P}_D$ is the acceleration due to the atmospheric drag and $\mathbf{P}_{SRP}$ is the acceleration due to the solar radiation pressure. The numerical integration of Equation 1 is made using the RADAU integrator [11].

The model of the acceleration due to the Geopotential used here can be found in Reference [12]. However, an important problem in the Geopotential model is related to the proper choice of the order and degree to be considered. It is chosen, for the Geopotential model used here, order and degree 8. Since the decay time is less than 25 years, and we are interested in the effects of solar radiation pressure and atmospheric drag, the quantity of harmonics in the Geopotential is acceptable. The calculation of the acceleration due to the atmospheric drag is given by[10]:

$$\mathbf{P}_D = -\frac{1}{2} C_D \, \rho \, \alpha \, |\mathbf{v}_r|^2 \frac{\mathbf{v}}{|\mathbf{v}|} \qquad (2)$$

where C_D is the aerodynamic coefficient. It is considered the value of 2.2. $\mathbf{v}_r$ is the relative velocity between the satellite and the atmosphere. $\mathbf{v}$ is the velocity of the satellite. α is the

[1] http://www.gaerospace.com/aeroassist/gossamer-orbit-lowering-device-gold-for-low-risk-satellite-de-orbit/

area-to-mass ratio. The minimum initial value (α_0) is equal to 0.007 m^2/kg. ρ is the atmospheric density. The paper considers the density up to 900 km altitude and used a table of medium density, which is made using the JB2008 atmospheric model. The calculation of the acceleration due to the solar radiation pressure is given by [10]:

$$\mathbf{P}_{SRP} = -P_{\odot}\, C_r\, \nu\, \alpha \,\frac{\mathbf{r}_{\odot}}{|\mathbf{r}_{\odot}|^3}\, AU^2 \tag{3}$$

where $P_{\odot}$ is the solar radiation pressure, determined by the solar flux. Because the distance between the Earth and the Sun is very large, $P_{\odot}$ is considered constant in the neighborhood of the Earth, so $P_{\odot} = 4.56 \times 10^{-6} Nm^{-2}$. It is worth mentioning that the area-to-mass has the same value used in the calculation of the acceleration due to the atmospheric drag. C_r is the reflectivity coefficient, AU is the Earth-Sun distance and ν is the shadow function. The shadow model used is found in Reference [10], and the eclipse factor takes into account the direct solar radiation, which has value 1, when totally exposed to the Sun, umbra, with value 0 and penumbra, with value between 0 and 1.

As mentioned before, this paper considers a satellite which carries a device that increases its area and which in also able to vary its reflectivity coefficient. As Figure 1 shows, C_r has a maximum value of 2 when $\langle \mathbf{P}_{SRP}, \mathbf{v} \rangle < 0$ and the minimum value of 1, in any other situation [6].

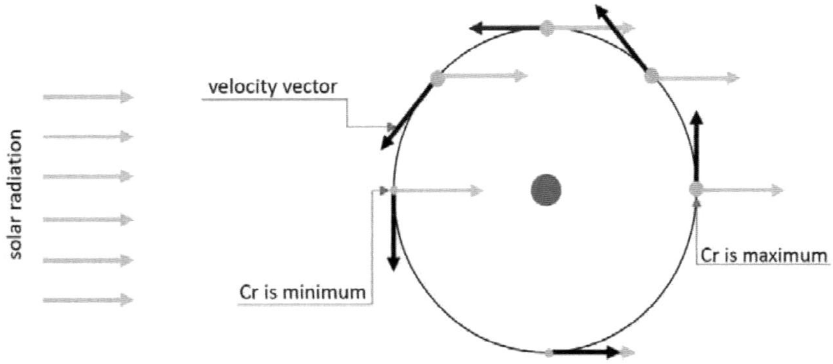

Figure 1. C_r variation control.

3. Results and Discussion

In this section, the results of the simulations are showed. In order to reach the highest altitudes possible, respecting the rule of 25 years, it is not considered the mission time to later trigger the device, i.e., the device has been active since the beginning of the mission. The initial conditions of the Sun and the satellite, common in all simulations, are presented in Table 1.

Table 1. Initial conditions for the Sun and the satellite.

	Sun	Satellite
Semi-major axis [km]	149512533.86795	-
Eccentricity	0.01617	0.005
Argument of perigee [degree]	284.245	0.0
Longitude of the ascending node [degree]	0.005	0.0
Mean anomaly [degree]	355.910	0.0
Inclination [degree]	23.437	30.0

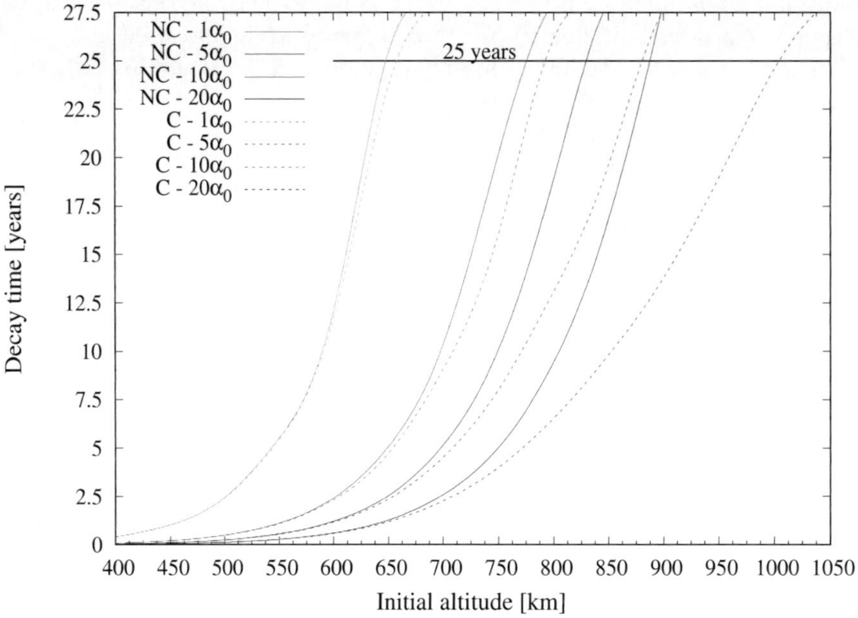

Figure 2. Cubesat decay time for the diferent situations.

Figure 2 presents the results for all simulations. The continuous curves (NC) show the results for the simulations where there is no presence of a control varying the coefficient of reflectivity. Its value is always equal to 1. The dashed curves (C) show the simulations where there is an active control varying the coefficient of reflectivity between 1 and 2, as previously mentioned. Simulations are done for different values of α_0, where: $\alpha = 1\alpha_0 = 0.007$ m^2/kg, $\alpha = 2\alpha_0 = 0.014$ m^2/kg, $\alpha = 5\alpha_0 = 0.035$ m^2/kg, and $\alpha = 20\alpha_0 = 0.14$ m^2/kg. As mentioned before, the presence of the atmosphere is considered up to 900 km. However, between 900 km and 700 km, the impact of the solar radiation pressure overcomes the effect of the atmospheric drag, which becomes dominant above 700 km [6]. We can see in Figure 2 that the difference between controlled and not controlled decays increases after 700 km of altitude. Moreover, the analysis of Figure 2 shows that:

- The decay times are near the same value up to 450 km of altitude for $5\alpha_0$, $10\alpha_0$ and $20\alpha_0$, and up to 500 km altitude for $10\alpha_0$ and $20\alpha_0$;
- In all cases, the decay time decreased as α increases. Ex.: The decays times for $\alpha = 5\alpha_0$, $\alpha = 10\alpha_0$ and $\alpha = 20\alpha_0$ at the altitude of 600 km show a decrease of approximately 10 years, 11.5 years and 12.25 years with respect to $\alpha = \alpha_0$;
- The decay times decrease as the value of α increases, from certain altitudes, when we compare the curves without control (NC cases) and with active control (C cases). Ex.: The decay times at altitude of 750 km show a decrease in the time of decay, for cases C with respect to NC cases: 4.8 years for $5\alpha_0$, 2.5 years for $10\alpha_0$ and 1.25 for $20\alpha_0$;
- The yellow curves obtained for α_0 show little difference in the times of decay up to approximately 600 km of altitude. The curves NC and C, with respect to the rule of 25 years, reach approximately 650 km and 662.5 km of altitude, respectively;
- The red curves obtained for $5\alpha_0$ show little difference in the times of decay up to approximately 600 km of altitude. The curves NC and C, with respect to the rule of 25 years, reach approximately 775 km and 795 km of altitude, respectively;

- The blue curves obtained for $10\alpha_0$ show little difference in the times of decay up to approximately 650 km of altitude. The curves NC and C, with respect to the rule of 25 years, reach approximately 830 km and 880 km of altitude, respectively;
- The green curves obtained for $20\alpha_0$ show little difference in the times of decay up to approximately 650 km of altitude. The curves NC and C, with respect to the rule of 25 years, reach approximately 830 km and 880 km of altitude, respectively.

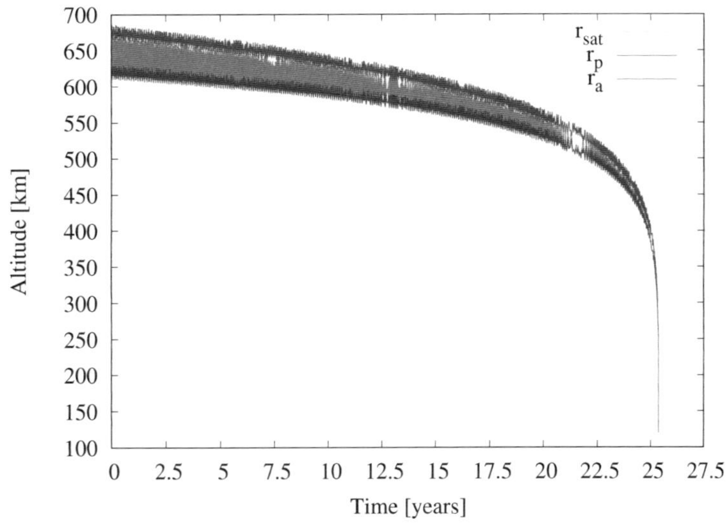

Figure 3. Decay of CubeSats from 650 km of altitude, without device and considering α=0.007 m^2/kg.

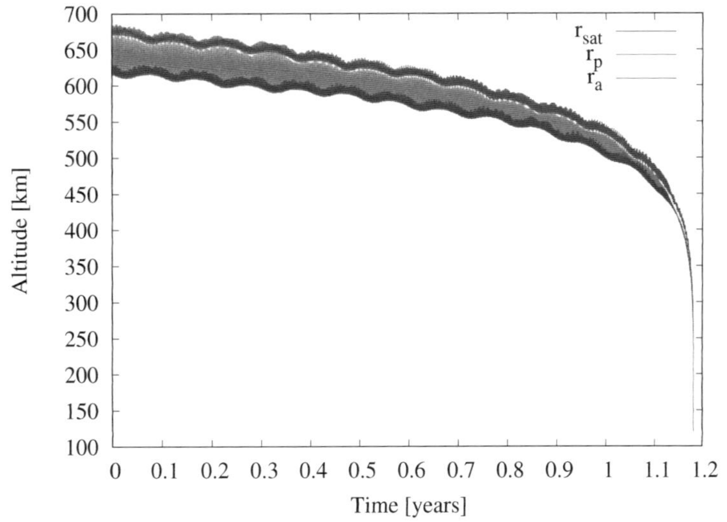

Figure 4. Decay of CubeSats from 650 km of altitude, with device and considering α=0.14 m^2/kg.

Figures 3 and 4 show the decay of a CubeSat from 650 km, where r_{sat} is the vector of the satellite position, r_p is the perigee radius and r_a is the apogee radius. Figure 3 shows the decay for the case without the device and considering $\alpha = 0.007$ m^2/kg. We can note that 650 km is near the initial altitude limit, so a CubSat with $\alpha = 0.007$ m^2/kg respect the rule of 25 years for reentry. Figure 4 shows the decay for the case with a device to increase the area, where $\alpha = 0.14$ m^2/kg, and able to vary the coefficient of reflectivity between 1 and 2, according to the control presented previously. When comparing Figure 3 with Figure 4, the advantage of using the proposed device is clear, where the decay time has dropped from approximately 25 years to approximately 1.18 years.

4. Conclusion

The present study shows that the area increase device and/or variation of the reflectivity coefficient increases the maximum altitude for missions where the rule of 25 years is respected and decrease the decay time for missions within the LEO region. The device of area increase and variation of the coefficient of reflectivity allows missions that respect the rule of 25 years where the initial altitude is in regions that the acceleration due to the atmosphere is negligible (above 900 km in our study), where the acceleration due to the radiation pressure being controlled is enough to cause the complete decay of the satellite. Future studies will make a broader analysis of the use of different coefficients of reflectivity and more area-to-mass settings, to try to increase even more the maximum altitude for missions of CubeSats, with respect the rule of the 25 years.

Acknowledgments: *the author thanks the grants # 406841/2016-0 and 301338/2016-7 from the National Council for Scientific and Technological Development (CNPq), and grants # 2014/22295-5, 2011/08171-3, 2016/14665-2, 2016/15675-1 and 2016/07248-6 from São Paulo Research Foundation (FAPESP).*

References

[1] Ereno D 2014 Pequenos ganham o espaço *Pesquisa FAPESP* **219** pp 16–23
[2] Hull S M 2011 *Space Mission Engineering: The New SMAD* ed Wertz J R, Everett D F and Puschell J J (Microcosm) pp 937–946
[3] IADC 2002 IADC space debris mitigation guidelines *Inter-Agency Space Debris Coordination Committee*
[4] Lücking C *et al* 2011 A passive de-orbiting strategy for high altitude CubeSat missions using a deployable reflective balloon *8th IAA Symp. on Small Satellites (Germany)*
[5] Guerman A and Smirnov G 2012 Orbital manoeuvres with single-input control *Advances in the Astronautical Sciences* **145** pp 171–181
[6] Deienno R *et al* 2016 Satellite de-orbiting via controlled solar radiation pressure *Celestial Mechanics and Dynamical Astronomy* pp 433–59.
[7] Silva Neto J B 2016 *Estudo da influência da pressão de radiação solar na remoção de satélites em ressonância* (São José dos Campos: INPE) p 132
[8] Qiao L 2013 Analysis and comparison of CubeSat lifetime *Proc. Australian Space Conf. (Australia)* pp 249–260
[9] Lokcu E *et al* 2011 A de-orbit system design for CubeSat payloads *Proc. Int. Conf. Rec. Ad. Space Tec.(Turkey)* pp 470–4
[10] Montenbruck O and Gill E 2001 *Satellite orbits* (Springer) p 369
[11] Everhart E 1985 An efficient integrator that uses Gauss-Radau spacings *Dynamics of comets: their origin and evolution* pp 185–202
[12] Sanchez D M 2014 On the effects of each term of the geopotential perturbation along the time I: quasi-circular orbits *Advances in Space Research* pp 1008–18

Mapping the Galilean moon's disturbance acting on a spacecraft's trajectory

Natasha Camargo de Araujo and Evandro Marconi Rocco

National Institute for Space Research, INPE, São José dos Campos, Brazil

natashac.araujo@gmail.com; evandro.rocco@inpe.br

Abstract. The prime objective of this work is to map the disturbance of Jupiter's Galilean moons, Io, Europa, Ganymede and Callisto, on a spacecraft trajectory. The study is done using an orbital trajectory simulator, the STRS (Spacecraft Trajectory Simulator). This mapping is made first considering the four moons as a group, and after that the disturbances of each of the Galilean moons are considered individually.

1. Introduction

This work aims to evaluate the disturbance of the four Galilean moons (Io, Europa, Ganymede and Callisto) on a spacecraft's trajectory orbiting Jupiter. These results were obtained by a short period simulation, of only 17 days, the same period of Callisto's orbit. It's important to emphasize that the goal of the study is not to analyze the dynamic evolution of the Jovian moons system for a long period of simulation, where the Laplace resonance must be considered due to its strong influence [2], but to consider a short period of time, a few days, equivalent to the orbital period of the farthest Galilean moon from Jupiter, to evaluate the magnitude of the Galilean moons gravitational disturbance applied in the spacecraft trajectory during this period of time. In this case, although, the resonance effect on Io, Europa and Ganymede orbits were considered, the changes in their orbits are not large in this short period of time. At the beginning of the simulation, the relative position of all moons was obtained from the Jet Propulsion Laboratory HORIZONS-Web interface [1]. From these data it is possible to propagate the relative position of the bodies during the whole simulation. However, when it comes to longer simulations, updating the relative positions using the HORIZONS-Web interface data is done more frequently in order to consider the effect of Laplace resonance.

2. Orbital trajectories simulator

The study of the Galilean moons disturbance acting on a spacecraft trajectory is done with the STRS (Spacecraft Trajectory Simulator), developed by [3, 4] and used by [5 - 10] at INPE. In the STRS, the orbital motion is obtained determining the spacecraft's state (position and velocity) at each step of the simulation. Moreover, the STRS propagates the relative positioning of all four Galilean moons in time [11], starting with the data obtained in the HORIZONS-Web interface, besides it considers the orbit propagation of other 58 Jupiter's moons.

3. Orbital disturbances

The Galilean moons disturbances are considered as disturbances generated by a third body individually, them the STRS sums all of them to ensure a complete simulation. The function of the gravitational potential due to the presence of a third body is given by equation 1 and can be found in Chobotov [12] and was used by [13] and [14].

$$\boldsymbol{F'} = \left(\frac{\mu'}{\boldsymbol{r'}}\right)\left[1 + \sum_{n=2}^{\infty}\left(\frac{r}{r'}\right)^n P_n \cos\psi\right] \tag{1}$$

μ is the gravitational parameter, $\boldsymbol{r'}$ is the absolute position vector related to the center of Jupiter, ψ is the angle between the spacecraft's position vector related to Jupiter and the spacecraft's position vector related with the third body, and $\boldsymbol{r}$ is the absolute position related to Jupiter. The disturbing accelerations due to a presence of a third body are shown from equation 2 to 4. According to [15] and [16].

$$\ddot{\boldsymbol{r}}_1 = -Gm_2\frac{\boldsymbol{r}_1 - \boldsymbol{r}_2}{|\boldsymbol{r}_1 - \boldsymbol{r}_2|^3} + Gm_3\frac{\boldsymbol{r}_3 - \boldsymbol{r}_1}{|\boldsymbol{r}_3 - \boldsymbol{r}_1|^3} \tag{2}$$

$$\ddot{\boldsymbol{r}}_2 = -Gm_3\frac{\boldsymbol{r}_2 - \boldsymbol{r}_3}{|\boldsymbol{r}_2 - \boldsymbol{r}_3|^3} + Gm_1\frac{\boldsymbol{r}_1 - \boldsymbol{r}_2}{|\boldsymbol{r}_1 - \boldsymbol{r}_2|^3} \tag{3}$$

$$\ddot{\boldsymbol{r}}_3 = -Gm_1\frac{\boldsymbol{r}_3 - \boldsymbol{r}_1}{|\boldsymbol{r}_3 - \boldsymbol{r}_1|^3} + Gm_2\frac{\boldsymbol{r}_2 - \boldsymbol{r}_3}{|\boldsymbol{r}_2 - \boldsymbol{r}_3|^3} \tag{4}$$

where $\ddot{\boldsymbol{r}}_1$, $\ddot{\boldsymbol{r}}_2$ and $\ddot{\boldsymbol{r}}_3$, are positions of the bodies of mass $\mathrm{m_1}$, $\mathrm{m_2}$ and $\mathrm{m_3}$ and G is the gravitational constant. At this point, it is important to explain that the three bodies involved here are Jupiter, the spacecraft and each of the Galilean moons, this calculus are done for all moons. The STRS computes the disturbing accelerations for each moon and sum all of them, considering, by this way, all of the four moons on results.

4. results

The orbital inclination and the semi-major axis were varied in a way to allow the mapping of the gravitational perturbation due to gravitational attraction of Io, Europa, Ganymedes and Callisto. In order to complete the mapping, the simulations have been done varying the inclination by 15°, from 0° to 90°, and the semi-major axis were also varied with values that could guarantee the spacecraft to always be positioned between two different Galilean moons. The 3D graphs and some of its projections are presented in figure 1 to 3.

The simulations were carried out using step of 1 minute, considering 17 terrestrial days as total time for the simulation, since Callisto's period, the farthest Galilean moon from Jupiter, is 16.69 days. It is important to point out that the simulations were performed to obtain the sum of the velocity increment due to the disturbance during this period of 17 days, for each combination of orbital inclination and semi-major axis.

To map the disturbing velocity increment, in each simulation a stable trajectory was considered only to determine what would be the sum of the disturbing velocity increment that a vehicle would be subjected on this trajectory. This sum can be equalized to the integral of the perturbation to which the vehicle would be subjected on this trajectory during the simulated period. The sum of the velocity increment is a characteristic of the chosen trajectory that could be considered in the mission analysis in order to obtain the minimally disturbed trajectory around Jupiter. Therefore mapping gravitational disturbances in this region is fundamental for the future missions toward Jupiter's system.

Some results are presented individually for each one of the Galilean moons (figures 4 to 15), however, the simulation of the spacecraft dynamics consider the gravitational attraction of all the four moons simultaneously. The results are presented in this way to make easier to understand how each moon contributes with the total gravitational disturbance applied to the spacecraft. However, figures 1 to 3 present the results for the sum of the disturbances due to all Galilean moons.

From figure 1 to 3, on the x axis is shown the semi-major axis, on the y axis the inclination and on z the velocity increment due the Galilean moons disturbances. As can be analyzed from the figures, the disturbances do not follow a rule when it comes to the semi-major axis, the greatest velocity increments happen when the spacecraft perform high approaches from the moons. However Fig. 3 shows that the lowest the inclination the higher the disturbances, this was expected as the Galilean moons have almost equatorial orbits.

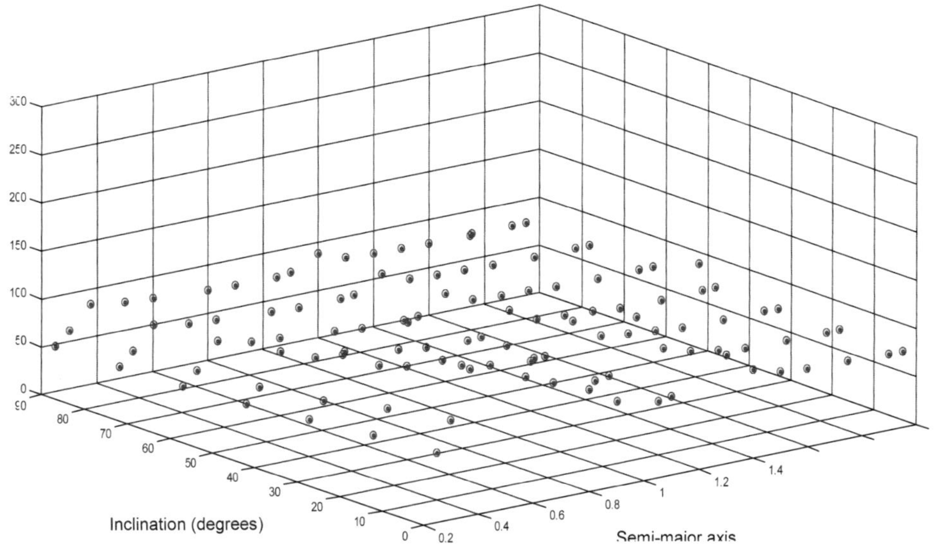

Figure 1. Galilean moons mapping.

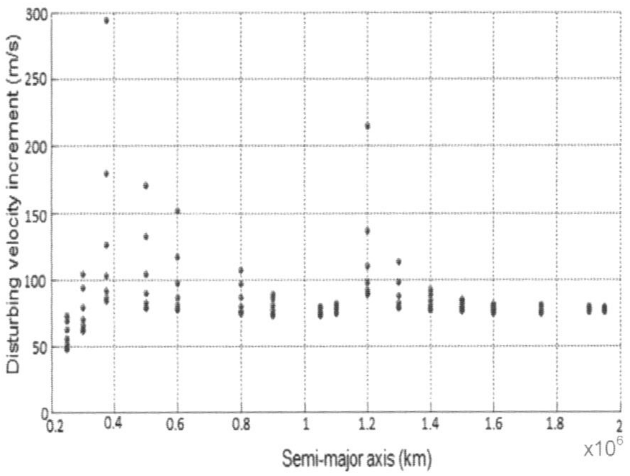

Figure 2. Galilean moons mapping (x-z projection).

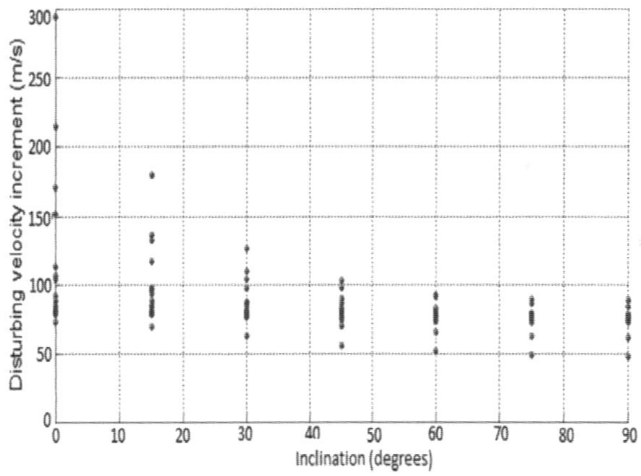

Figure 3. Galilean moons mapping (y-z projection).

Now the same graphs are shown for each of the Galilean moons, Io, Europa, Ganymede and Callisto from figure 4 to 15.

As can been seen in figure 4 to 6, for Io, the disturbances are greater for smaller semi-major axis, that's when the spacecraft approaches to the moon, and when it comes to the inclination, again as analyzed for the Galilean moons as a group the more equatorial the orbits, the greater the disturbances. For the other moons, from figure 7 to figure 15, Io's pattern keeps repeating, when the spacecraft perform great approaches, the velocity increment increases, and for all the Galilean moons as a group or individually when the inclination is small there are greater disturbances.

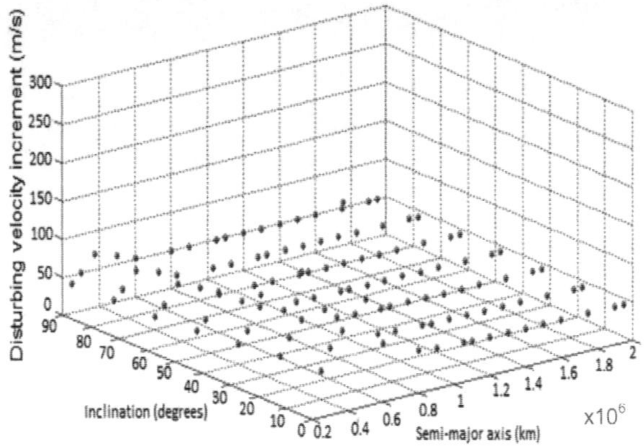

Figure 4. Mapping of Io's disturbances.

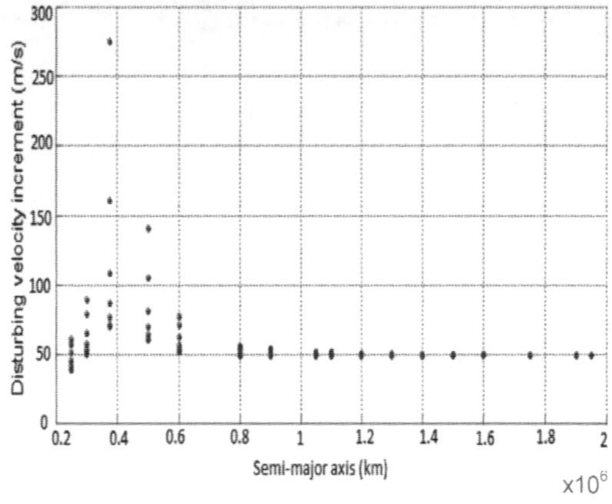

Figure 5. Mapping of Io's disturbances (x-z projection).

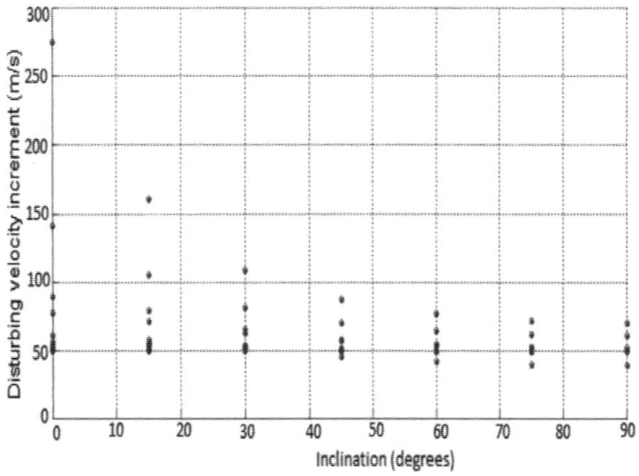

Figure 6. Mapping of Io's disturbances (y-z projection).

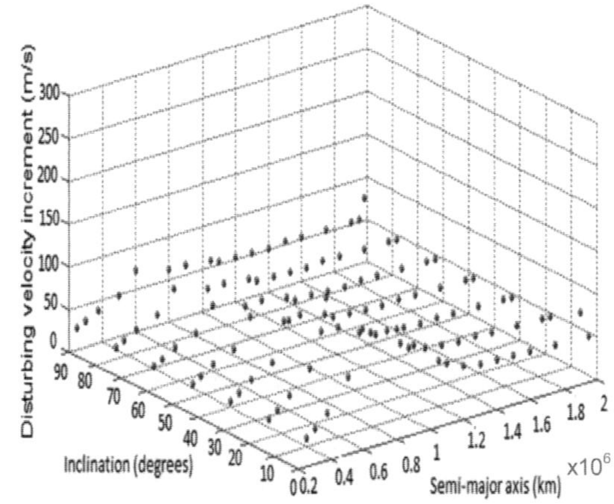

Figure 7. Mapping of Europa's disturbances.

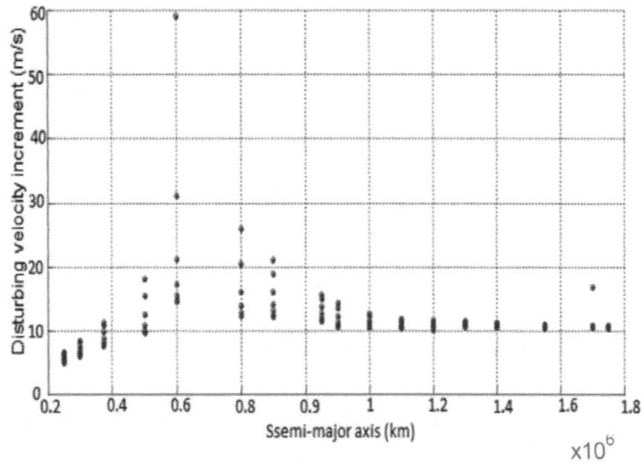

Figure 8. Mapping of Europa's disturbances (x-z projection).

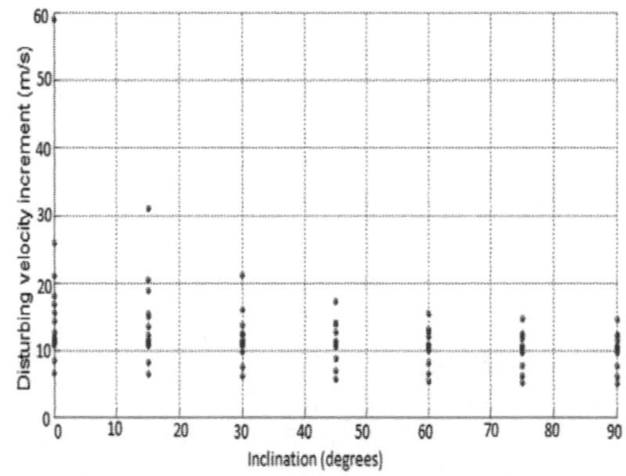

Figure 9. Mapping of Europa's disturbances (y-z projection).

IOP Conf. Series: Journal of Physics: Conf. Series **911** (2017) 012010 doi:10.1088/1742-6596/911/1/012010

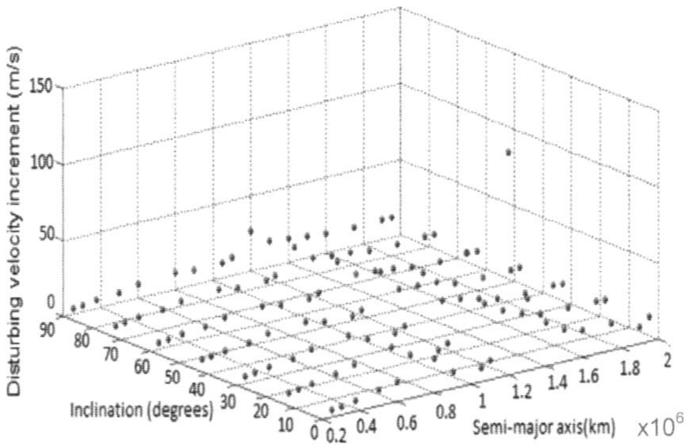

Figure 10. Mapping of Ganymede's disturbances.

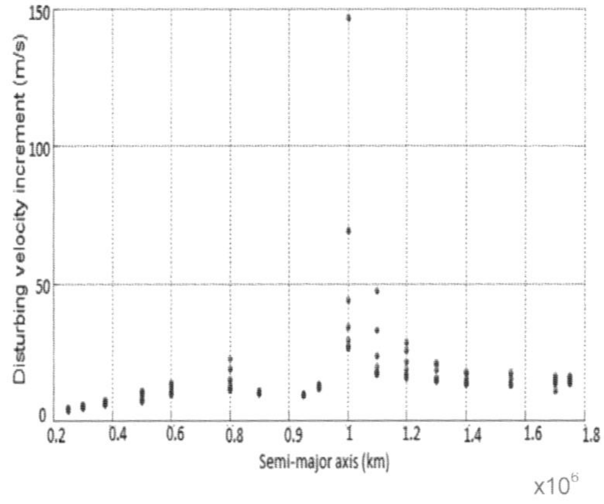

Figure 11. Mapping of Ganymede's disturbances (x-z projection).

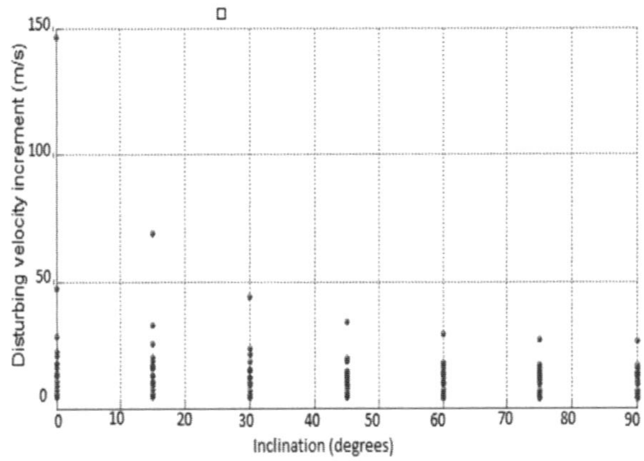

Figure 12. Mapping of Ganymede's disturbances (y-z projection).

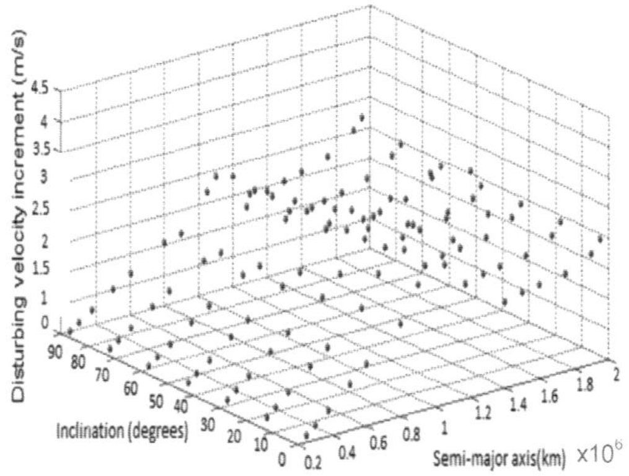

Figure 13. Mapping of Callisto's disturbances.

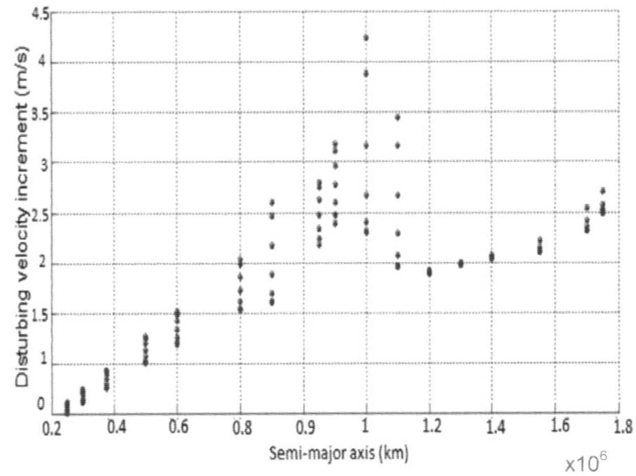

Figure 14. Mapping of Callisto's disturbances (x-z projection).

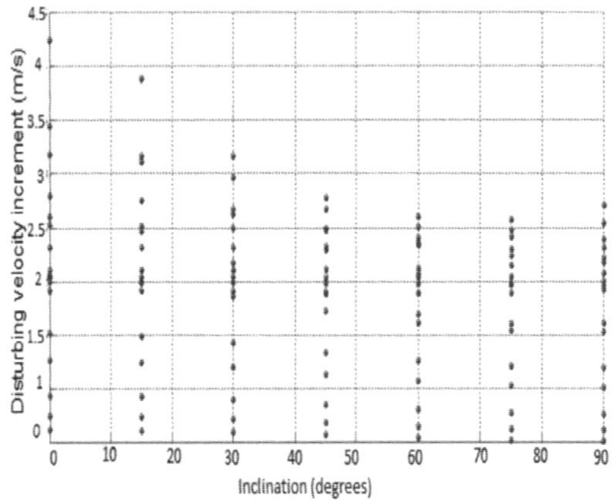

Figure 15. Mapping of Callisto's disturbances (y-z projection).

5. Conclusion

The Galilean moons, as a group or individually, have a large influence on the spacecraft's trajectory. The velocity increments generated by their disturbances are not small, and should be considered when it comes to missions involving Jupiter. As these disturbances are a sum, they are greater as time increases, and generally this type of mission lasts some years, so these disturbances, that are already great, affect more and more the trajectories through time. By this way, it is evident that the disturbances on trajectory owing to the gravitational attraction of the Galilean moons cannot be neglected in mission analysis of a space vehicle that aims to orbit Jupiter. Thus the disturbance models due to the Galilean moons were considered in the trajectory simulator STRS and simulations were successfully performed showing the disturbance velocity increments caused by these moons.

References

[1] HORIZONS-Web interface-Jet Propulsion Laboratory (NASA). California Institute of Technology (Caltec). [Cited 2016 Feb 10]. Available from: https://ssd.jpl.nasa.gov/horizons.cgi

[2] Sinclair A T 1975 *The Orbital Resonance Amongst the Galilean Satellites of Jupiter. Monthly Notices of the Royal Astronomical Society.* ***171*** *(1):59-72*

[3] Rocco E M 2008 *Perturbed orbital motion with a PID control system for the trajectory* (Águas de Lindóia: XIV Colóquio Brasileiro de Dinâmica Orbital) pp 17-21

[4] Rocco E M 2012 *Controle de trajetória com propulsão contínua para missões do tipo drag-free* (Maranhão: Congresso Nacional de Engenharia Mecância)

[5] Rocco E M 2013 *Automatic correction of orbital elements using continuous thrust controlled in closed loop.* Journal of Physics. Conference Series, v. 465, p. 012007. doi:10.1088/1742-6596/465/1/012007

[6] dos Santos W G; Kuga H K; Rocco E M 2013 *Application of the Kalman Filter to Estimate the State of an Aero braking Maneuver.* Mathematical Problems in Engineering, v. 2013, p. 1-8. doi.org/10.1155/2013/234270

[7] dos Santos W G; Rocco E M; Carrara V 2014 *Trajectory Control During an Aeroassisted Maneuver Between Coplanar Circular Orbits.* Journal of Aerospace Technology and Management, v. 6, p. 159-168. doi: 10.5018/jatm.v6i2.351

[8] Oliveira T C; Rocco, E M; Prado, A F B A; Ferreira, J L 2013 *A Study of the Duration of the Passage through the Van Allen Belts for a Spacecraft going to the Moon.* Journal of Physics. Conference Series, v. 465, p. 012019. doi:10.1088/1742-6596/465/1/012019

[9] Venditti, F C F; Rocco, E M; Prado, A F B A 2013 *Trajectory control around non-spherical bodies modelled by parallelepipeds.* Journal of Physics. Conference Series, v. 465, p. 012008. doi:10.1088/1742-6596/465/1/012008

[10] Gonçalves L D; Rocco E M; de Moraes, R V 2013 *Orbital Disturbance Analysis due to the Lunar Gravitational Potential and Deviation Minimization through the Trajectory Control in Closed Loop.* Journal of Physics. Conference Series, v. 465, p. 012013. doi:10.1088/1742-6596/465/1/012013

[11] Rocco E M 2015 *Maneuvers around Jupiter with automatic correction of the trajectory considering gravitational disturbances generated by the Galilean moons: IO, Europa Ganymede and Callisto.* In: 25th International Symposium on Space Flight Dynamics, Conference Proceedings of 25th International Symposium on Space Flight Dynamics, v. 1.

[12] Chobotov V A 2002 *Orbital mechanics* (American Institute of Aeronautics and Astronautics)

[13] Gonçalves L D; Rocco E M; de Moraes R V 2015 *Analysis of the influence of orbital disturbances applied to an artificial lunar satellite.* Journal of Physics. Conference Series, v. 641, p. 012028. doi:10.1088/1742-6596/641/1/012028

[14] Rocco E M 2015 *Gravitational disturbances generated by the Sun, Phobos and Deimos in orbital maneuvers around Mars with automatic correction of the semi-major axis.* Journal of Physics. Conference Series, v. 641, p. 012027. doi:10.1088/1742-6596/641/1/012027

[15] Szebehely 1967 *Theory of orbits* (New York: Academic Press Inc.)

[16] Prado A F B A and Kuga H K 2001 *Fundamentos de tecnologia espacial* (INPE)

IOP Conf. Series: Journal of Physics: Conf. Series **911** (2017) 012011 doi:10.1088/1742-6596/911/1/012011

Analysis of the long and short-period terms due the nonsphericity of the central body: applications for Mercury

J. P. S. Carvalho,[1]

[1] UFRB-Centro de Ciência e Tecnologia em Energia e Sustentabilidade, Universidade Federal do Recôcavo da Bahia, Feira de Santana-BA, Brazil

E-mail: jeanfeg@gmail.com

Abstract.
In this work, we present an approach taking into account the single-averaged equations and unaveraged equations to investigate the dynamics of artificial satellites on the effect due to the non-spherical shape of the planet Mercury. An analysis considering the long-period terms and another taking into account the short-period terms is presented. The numerical integrations of the equations developed are performed using the Maple software. We consider the numerical values of the most updated spherical harmonic coefficients in the literature. Emphasis is given to analyze the effect of the C_{22} term in the dynamics of the spacecraft. We show that the two techniques are in agreement (average or not average). We found orbits that librates around an equilibrium point with small variation of the orbital elements, in particular the eccentricity and argument of the pericenter. We also note that the C_{22} term contributes to reduce the growth of the orbital eccentricity.

1. Introduction

Several authors have analyzed the dynamics of artificial satellites around Mercury, taking into account the non-spherical shape of the planet, see for example, [1, 2, 3, 4, 5]. These authors have considered the values of the harmonic coefficients obtained by the Mariner probe (1974). But, these coefficients are now available in the literature in an updated version, see for example, [6, 7]. In [6] it is realized an analyze 3 years of radio tracking data from the MESSENGER spacecraft in orbit around Mercury, where the gravity field of the planet was determined. Here, in this research, we using the updated values of the spherical harmonics coefficients to describe the gravity field of Mercury, obtained from [6]. In this work, an approach that takes into account the effect due to the non-spherical shape of the planet Mercury on the dynamics of artificial satellites is presented, where an analysis considering the terms of long-period and another taking into account the terms of short-period is considered .

2. Nonsphericity of the central body

In this section, are presented the long-period equations, where the averaged model is applied. In the next section, we show the development of the short-period equations. Consider the satellite orbit around the central body with semimajor axis a, eccentricity e, inclination i, right ascension of the ascending node h, argument of the pericentre g and mean motion n.

IOP Conf. Series: Journal of Physics: Conf. Series **911** (2017) 012011 doi:10.1088/1742-6596/911/1/012011

To develop the disturbing potential, that represents the non-uniform distribution of mass of the planet, we use Eq. (1)(see [8])

$$U_M = -\frac{\mu}{r}\left[\sum_{n=2}^{\infty}\left(\frac{R_e}{r}\right)^n J_n P_n(\sin\phi) - \left(\frac{R_p}{r}\right)^2 C_{22}P_{22}(\sin\phi)\cos(2\lambda)\right], \quad (1)$$

where μ and R_e are the gravitational parameter and equatorial radius of the planet, respectively. Here P_n are the Legendre polynomials, P_{nm} the associated Legendre polynomials, the angle ϕ is the latitude of the orbit with respect to the equator of the planet, the angle λ is the longitude measured from the direction of the longest axis of the planet. However, λ contains the time explicitly (see [8]). Where r is the position of the satellite with respect to the Mercury center.

The disturbing potential due to the non-uniform distribution of mass of the central body up to the third order for the zonal terms and the C_{22} term (equatorial ellipticity) it is presented in [9, 10, 11]. We get,

$$< R_{J2} >= -\frac{1}{4}\frac{J_2 R_e^2}{(1-e^2)^{3/2}} n^2 (3s^2 - 2) \quad (2)$$

$$< R_{J3} >= -\frac{3}{8}\frac{J_3 R_e^3}{(1-e^2)^{5/2} a} n^2 se\sin(g)(5s^2 - 4) \quad (3)$$

$$< R_{C_{22}} >= -\frac{3}{2}\frac{C_{22} R_e^2}{(1-e^2)^{3/2}} n^2 (c^2 - 1)\cos(2wt - 2h) \quad (4)$$

where $s = \sin i$ and $c = \cos i$. Here $w = 6.138506839°/day$ is the rotation rate of Mercury ([12]) and t is the time.

Thus, the disturbing potential due to the effect of the non-spherical shape of the central body can be written as follows

$$R =< R_{J_2} > + < R_{J_3} > + < R_{C_{22}} > \quad (5)$$

Considering the equations that represent the non-sphericity of the central bodoy, the artificial satellite is studied under the single-averaged analytical model. The coefficients of the gravity field is given in table 1. Note that in Eq. (4) the time appears explicitly on the right side. Replacing the potential given by Eq.(5) in the Lagrange planetary equations we obtain a system of non-linear and non-autonomous differential equations. This system of differential equations is numerically integrated using the Maple software.

Table 1. Normalized harmonics coefficients of Mercury ([6]).

Coefficients	Value
J_2	2.25045×10^{-5}
J_3	4.76589×10^{-6}
C_{22}	1.24538×10^{-5}

IOP Conf. Series: Journal of Physics: Conf. Series **911** (2017) 012011 doi:10.1088/1742-6596/911/1/012011

2.1. Short-period equations

Now, the development of the short-period equations is presented, were the averaged model is not applied. The Legendre polynomials for the J_2 and J_3 zonal terms and the Legendre associated functions for the C_{22} sectorial term can be written in the form [8]

$$\begin{aligned} &P_2(\sin\phi) = \tfrac{1}{2}(3s^2\sin^2(f+g)-1),\\ &P_3(\sin\phi) = \tfrac{5}{2}(s^3\sin^3(f+g)) - \tfrac{3}{2}s\sin(f+g),\\ &P_{22}(\sin\phi)\cos 2\lambda = 6(\xi^2\cos^2 f + \chi^2\sin^2 f + \xi\chi\sin 2f) - 3(1-s^2\sin^2(f+g)), \end{aligned} \tag{6}$$

where we used the shortcut $\xi = \cos g\cos h - c\sin g\sin h$ and $\chi = -\sin g\cos h - c\cos g\sin h$.

To represent the flattening due to J_2 term, Eqs. (1) and (6) are used. We get,

$$U_{20} = -\tfrac{1}{2}\tfrac{a^3}{r^3}\epsilon n^2(3s^2(\sin^2(f+g)) - 1) \tag{7}$$

where $\epsilon = J_2R_e^2$.

We used known equations of celestial mechanics to put the equation in function of the orbital elements (see [13]), we get

$$a/r = (1+e\cos(f))/(1-e^2) \tag{8}$$

$$\cos(f) = \cos(l) + e(\cos(2l)-1) + e^2\left(\tfrac{9}{8}\cos(3l) - \tfrac{9}{8}\cos(l)\right) \tag{9}$$

$$\sin(f) = \sin(l) + e\sin(2l) + e^2\left(\tfrac{9}{8}\sin(3l) - \tfrac{7}{8}\sin(l)\right) \tag{10}$$

where l is the mean anomaly and f is true anomaly of the satellite. The expression $\mu = n^2a^3$ is also used.

Using Eqs. (7), (8), (9) and (10) and after some algebraic manipulations, the potential due to J_2 term is written as follows

$$\begin{aligned} R_{J_2} = &-\tfrac{15}{8}\epsilon((e^2-2/5)(s)^2\cos(2g+2l) - 7/5e(s)^2\cos(2g+3l) -\\ &\tfrac{17}{5}e^2(s)^2\cos(2g+4l) + 1/5e(s)^2\cos(l+2g) +\\ &9/5((s)^2-2/3)(e^2\cos(2l) + 1/3e^2 + 2/3e\cos(l) + 2/9))n^2 \end{aligned} \tag{11}$$

To represent the flattening due to J_3 term, Eqs. (1) and (6) are used. We get,

$$U_{30} = -\tfrac{1}{2}\tfrac{a^4}{r^4}\tfrac{n^2}{a}\epsilon_1(5s^3(\sin(f+g))^3 - 3s\sin(f+g)) \tag{12}$$

$\epsilon_1 = J_3R_e^3$.

Using Eqs. (12), (8), (9) and (10) and after some algebraic manipulations, the potential due to J_3 term is written as follows

$$\begin{aligned} R_{J_3} = &-\tfrac{165}{64}n^2\epsilon_1 s(\tfrac{16}{11}(e^2-1/6)(s)^2\sin(3g+3l) + \tfrac{8}{33}(s)^2\sin(2l+3g)e -\\ &\tfrac{40}{33}(s)^2\sin(4l+3g)e - \tfrac{127}{33}(s)^2\sin(5l+3g)e^2 + e^2((s)^2-4/5)\sin(g-l) +\\ &\tfrac{53}{11}e^2((s)^2-4/5)\sin(g+3l) + (\tfrac{24}{11}e(s)^2 - \tfrac{96}{55}e)\sin(2l+g) -\\ &1/33(s)^2\sin(l+3g)e^2 + \tfrac{16}{11}((e^2+1/2)\sin(g+l) + 1/2\sin(g)e)((s)^2-4/5))a^{-1} \end{aligned} \tag{13}$$

To describe the equatorial ellipticity, where we represent by the C_{22} term, we use Eq.(5) of the paper from De Saedeleer ([14]) given by Eq. (14)

$$U_{22} = \frac{3}{4}\frac{\delta\, n^2 a^3}{r^3}\left(2\, s^2 \cos(2\, h) + (c+1)^2 \cos(2\, f + 2\, g + 2\, h) + (c-1)^2 \cos(2\, f + 2\, g - 2\, h)\right) \tag{14}$$

but here $\delta = C_{22} R_e^2$.

Using Eqs. (14), (8), (9) and (10) and after some algebraic manipulations, the potential due to C_{22} term is written as follows

$$\begin{aligned} R_{C_{22}} = {}& \tfrac{9}{4}\delta n^2(-5/6(c-1)^2(e^2-2/5)\cos(2g+2l-2h)- \\ & 5/6(e^2-2/5)(c+1)^2\cos(2g+2l+2h) + 7/6e(c-1)^2\cos(2g+3l-2h)+ \\ & 7/6e(c+1)^2\cos(2g+3l+2h) + \tfrac{17}{6}e^2(c-1)^2\cos(2g+4l-2h)+ \\ & \tfrac{17}{6}e^2(c+1)^2\cos(2g+4l+2h) - 1/6e(c-1)^2\cos(2g+l-2h)- \\ & 1/6e(c+1)^2\cos(2g+l+2h) + (e^2\cos(2h) + 3/2e^2\cos(-2l+2h)+ \\ & 3/2e^2\cos(2l+2h) + e\cos(-l+2h) + e\cos(l+2h) + 2/3\cos(2h))(s)^2) \end{aligned} \tag{15}$$

For the time to appear explicitly in Eq. (15), we use the expression $h = h_0 - wt$, given by Sehnal ([15]), see also [8, 16]. Therefore, the longitude of the ascending node h in Eq. (15) should be replaced by $h = h_0 - wt$. For more details see the References [10, 16].

Thus, the short-period disturbing potential can be written in the form

$$K = R_{J_2} + R_{J_3} + R_{C_{22}} \tag{16}$$

3. Results

Now, the short-period perturbing potential (Eq. (16)) is replaced in the Lagrange planetary equations and numerically integrated using the Maple software. This section presents the results obtained from the numerical simulations. The numerical integration used the routine "dsolve" of the software Maple (version 17), with the options "numeric" and "method=rkf45". It finds a numerical solution using a Fehlberg fourth-fifth order Runge-Kutta method with degree four. Figures 1 and 2 show a comparison when the C_{22} term is considered in the dynamics (see Fig. 2) and when this term is neglected (see Fig. 1). Note that the C_{22} term contributes to reduce eccentricity growth, in particular, for the eccentricity 0.02.

For the single-averaged potential, given by Eq. (5), we obtain the following results. The diagram e x g, see Fig. 3, shows orbits that librates around an equilibrium point, see in particular the orbit with $e = 0.02$, and orbits that circulates ($e = 0.05, e = 0.06$). We call here the orbit with $e = 0.02$ as an almost frozen orbit, which presents a smaller variation of the orbital elements. Note that Fig. 2, considering the short-period terms, shows exactly this value of the eccentricity as the orbit with smaller growth. Figure 4 shows the behavior of the eccentricity for different initial values, as expected, the orbit with $e = 0.02$ has smaller growth when compared to the other trajectories. The orbit that circulates in Fig. 3 ($e = 0.05$), also as expected, shows a high rate of variation of the orbital eccentricity (see Fig. 4). The orbits shown in Figs. 3 and 4 are polar ($i = 90$ degrees).

Figures 5 and 6 show orbits with initial inclination of 65 degrees and 125 degrees, respectively. Note that more orbits circulate, and the orbits that librates around an equilibrium point, now show lager variation of the orbital elements when compared to Fig. 3. This characteristic of the polar orbit is important, as scientific missions often require polar orbits. For example, the BepiColombo is Europe's first mission to Mercury. It will set off in 2018. In this mission are planned two spacecraft orbiting Mercury both with inclination of 90 degrees.

IOP Conf. Series: Journal of Physics: Conf. Series **911** (2017) 012011 doi:10.1088/1742-6596/911/1/012011

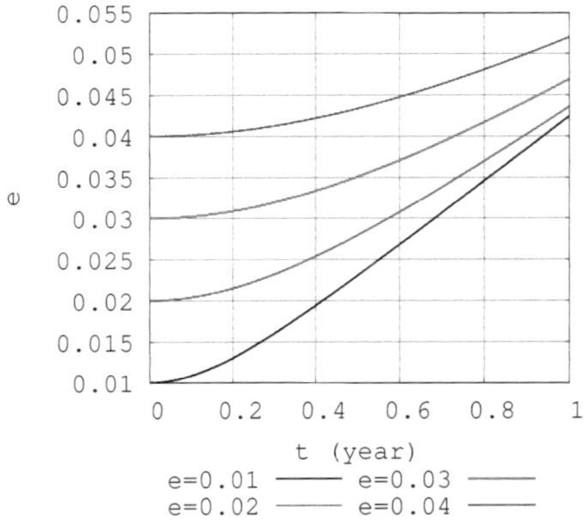

Figure 1. Initial conditions: $a = 2700$ km, $i = 90°$, $l = 0°$. $g = 270°$ and $h = 90°$. Disturbing Potential: $R_{J_2} + R_{J_3}$.

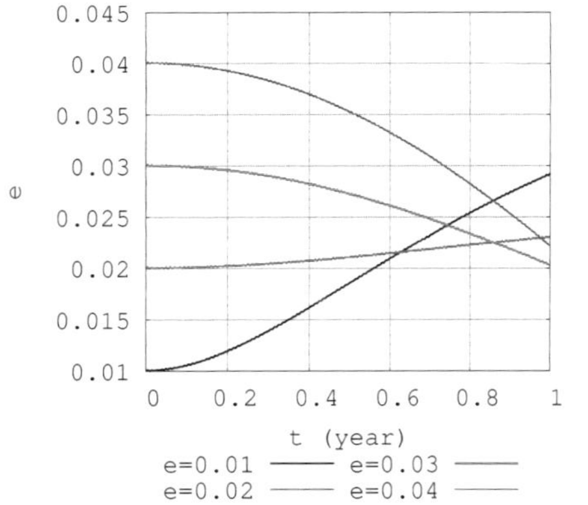

Figure 2. Initial conditions: $a = 2700$ km, $i = 90°$, $l = 0°$. $g = 270°$ and $h = 90°$. Disturbing Potential: $R_{J_2} + R_{J_3} + R_{C_{22}}$.

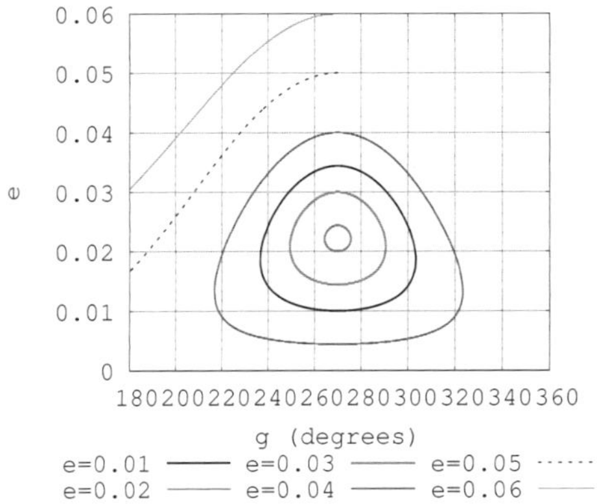

Figure 3. Diagram e versus g. Initial conditions: $a = 2700$ km, $i = 90°$, $g = 270°$ and $h = 90°$. Disturbing Potential: $R_{J_2} + R_{J_3} + R_{C_{22}}$.

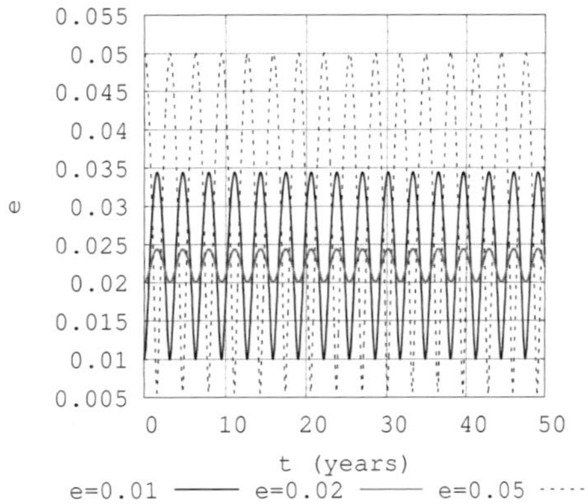

Figure 4. e versus t. Initial conditions: $a = 2700$ km, $i = 90°$, $g = 270°$ and $h = 90°$. Disturbing Potential: $R_{J_2} + R_{J_3} + R_{C_{22}}$.

4. Conclusions

An analytical approach is presented considering the single-averaged equations, and equations without averaged. These equations are numerically integrated to analyze the dynamics of artificial satellites around Mercury. The perturbation considered in this research is the non-uniform distribution of mass of the planet considering zonal terms and the C_{22} sectoral term. We show that the single-averaged model and the model without average are in agreement. In both models we found the orbit with smaller variation of the orbital elements, in particular the eccentricity. In previous work we have already shown that the C_{22} term coupled with the zonal

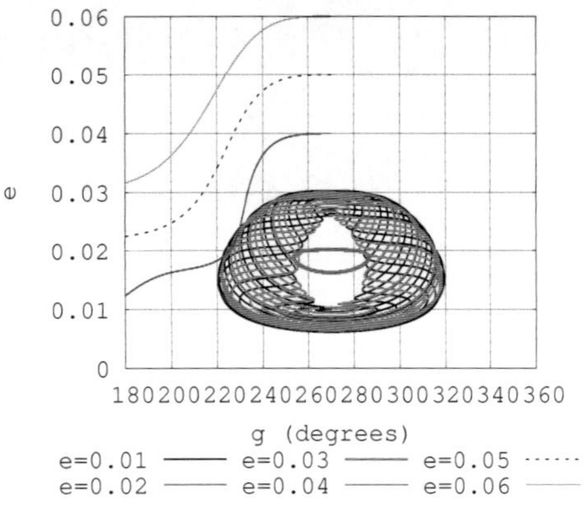

Figure 5. Diagram e versus g. Initial conditions: $a = 2700$ km, $i = 65°$. $g = 270°$ and $h = 90°$. Disturbing Potential: $R_{J_2} + R_{J_3} + R_{C_{22}}$.

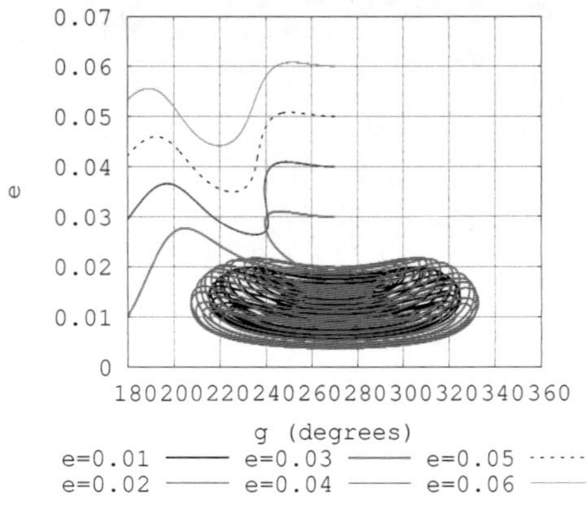

Figure 6. Diagram e versus g. Initial conditions: $a = 2700$ km, $i = 125°$. $g = 270°$ and $h = 90°$. Disturbing Potential: $R_{J_2} + R_{J_3} + R_{C_{22}}$.

terms helps to control the growth of the eccentricity, here we also show that the same happens with the equations without average.

5. Acknowledgments
Sponsored by CNPq - Brazil. The authors are grateful to CNPq (National Council for Scientific and Technological Development) - Brazil for contracts 306953/2014-5 and 420674/2016-0.

References
[1] Lara, M., Palacián, J. F., Yanguas, P., Corral, C., 2010 *Acta Astronautica* **66** 1022
[2] Ma, X., Li, J., 2013 *Astrophys Space Sci* **348** 345
[3] Tresaco, E., Eipe, A., Carvalho, J. P. S., 2016 *J. Guid. Control. Dynam.* **39**(7)1659
[4] Delsate, N., Robutel, P., Lemaître, A., Carletti, T., 2010 *Celest. Mech. Dyn. Astron.* **108** 275
[5] Carvalho, J. P. S., 2016 *Proceeding Series of the Brazilian Society of Applied and Computational Mathematics,* **4**(1) 010017-1
[6] Mazarico, E., Genova, A., Goossens S., et al., 2014 *J. Geophys. Res. Planets,* **119** 2417
[7] Verma, A. K., Margot, J. -L., 2016 *J. Geophys. Res. Planets* **121** 1
[8] Giacaglia, G.E.O., Murphy J., Felsentreger T., 1970 *Celest. Mech.* **1**, 3, 3
[9] Carvalho, J. P. S., Vilhena de Moraes, R., Prado, A. F. B. A., 2013 *Math. Probl. Eng.*, 2013, Article ID 182079
[10] Tresaco, E., Carvalho, J. P. S., Prado, A. F. B. A., Elipe, A., Vilhena de Moraes, R. "To appear", 2017.
[11] Carvalho, J. P. S., Tresaco, E., Vilhena de Moraes, R., Prado, A. F. B. A., 2016 *XXXVI Congresso Nacional de Matemtica Aplicada e Computacional* - CNMAC2016, Gramado-RS. Proceeding Series of the Brazilian Society of Computational and Applied Mathematics, 1
[12] Stark, A. Oberst, J., Hussmann H., 2015 *Celest. Mech. Dyn. Astron.*, **123** 263
[13] Murray, C. D., Dermott, S. F. Solar System Dynamics. Cambridge: Cambridge University Press, 1999.
[14] De Saedeleer, B., 2006 *Celest. Mech. Dyn. Astron.* **95** 407
[15] Sehnal, L., 1959, *Bulletin of the Astronomical Institute of the Czechoslovak Academy of Sciences,* **11** 90
[16] Meyer, W. K., Buglia, J. J., Dsai, P. N., 1994 *NASA STI/Recon Technical Report N-TP-3394*, **94**, 27771

Injection of a microsatellite in circular orbits using a three-stage launch vehicle

L O Marchi[1], J O Murcia[1], A F B A Prado[1] and C R H Solórzano[2]

[1] National Institute for Space Research – INPE, São José dos Campos (SP), Brazil

[2] Federal University of ABC – UFABC, Santo André (SP), Brazil

E-mail: marchi.luis@yahoo.com.br

Abstract. The injection of a satellite into orbit is usually done by a multi-stage launch vehicle. Nowadays, the space market demonstrates a strong tendency towards the use of smaller satellites, because the miniaturization of the systems improve the cost/benefit of a mission. A study to evaluate the capacity of the Brazilian Microsatellite Launch Vehicle (VLM) to inject payloads into Low Earth Orbits is presented in this paper. All launches are selected to be made to the east side of the Alcântara Launch Center (CLA). The dynamical model to calculate the trajectory consists of the three degrees of freedom (3DOF) associated with the translational movement of the rocket. Several simulations are performed according to a set of restrictions imposed to the flight. The altitude reached in the separation of the second stage, the altitude and velocity of injection, the flight path angle at the moment of the activation of the third stage and the duration of the ballistic flight are presented as a function of the payload carried.

1. Introduction

According to the Brazilian "National Space Activities Program" (PNAE), it is necessary the development of critical technologies in the Brazilian space industry [1]. During the design process and the launch of space rockets, it is required modelling and simulations. Recently, researchers at INPE and ITA presented a computational tool to simulate the motion of a launch vehicle in 3DOF and 6DOF [2,3]. Another studies explored various engine configurations for rocket liquid propellant and to calculate the performance of the Satellite Launch Vehicle – VLS [4]. The study to analyse the dynamics of the separation of the first stage of the VLM to quantify the displacements and rotations between the vehicle and the discarded stage was performed in 2015 [5]. The problem of the VLM trajectory optimization is performed by [2,6]. The main objective of this work is to make a preliminary analysis on the capacity of the VLM to inject payloads in circular orbits with inclination close to 2,3°.

2. Methodology

The Brazilian Microsatellite Launch Vehicle (VLM) is selected as the launch vehicle to perform the simulations. It is a vehicle with approximately 28 ton composed of three stages of solid propulsion configured in series (tandem) [6].

The launch site is Alcântara (CLA), with geodetic coordinates 2°22'39.52''S, 44°23'57.71''W at the sea medium level. The atmospheric model is the US ATM 1976, rotating with the planet and without winds [7]. The mathematical expansion of the Earth's geopotential model up to the J_4 spherical harmonic was performed [8]. However, in this work it was considered up to the J_6 spherical harmonic due to the proximity of Alcântara with the terrestrial Equator. The rocket engine is ideal, the thrust force is associated to the propellant specific impulse, without variations in thrust due to the atmospheric pressure. A Runge-Kutta-Fehlberg seventh order (RKF-7) numerical integrator with fixed stepsize (10^{-3}) solves the differential equations [9]. The VLM technical data are presented in Table 1.

Table 1. VLM's data [6].

PARAMETERS	UNITS	STAGE I	STAGE II	STAGE III	FAIRING
Diameter	cm	135	135	101	120
Liftoff weight	kg	12900	13100	1004	110
Propellant weight	kg	11500	11500	814	-
Specific impulse	s	286.4	286.4	289	-
Burn time	s	85	85	71	-

3. Formulation of the dynamical model

To determine the possible flight path of the VLM, it is necessary to analyze the initial conditions on the launch site and the forces applied in its center of gravity along the trajectory. Three forces are taking into account in the mass-point body or carrier rocket. The principal equation that describes the dynamical system is Equation 1, where $\ddot{r}$ is the inertial acceleration.

$$\ddot{\vec{r}} = \frac{\vec{T}}{m} - \frac{\vec{D}}{m} - \vec{g} \tag{1}$$

where $\vec{T}$ is the force coming from the engine, $\vec{D}$ the aerodynamic drag, $\vec{g}$ the gravity and m is the mass of the body, which changes as a function of time due to the mass flow consumption.

The force opposite to the motion of the spacecraft, known as drag, is proportional to the projected area of the body (A), the drag coefficient (C_D), the atmospheric density (ρ) and the velocity of the spacecraft with respect to the atmosphere (V).

$$D = \frac{1}{2}\rho A C_D V^2 \tag{2}$$

The description of the equations in the local reference system and the transformation between different reference systems are available in [10].

The dynamic equations in 3DOF are integrated in the Earth-Centered, Earth-Fixed (ECEF) system. The inertial reference system (ECI) is used to calculate the inclination of the injection orbit. The trajectory is guided through the local navigation system (ENU) whose origin is the center of mass of the rocket. These three coordinate systems are shown in Figure 1.

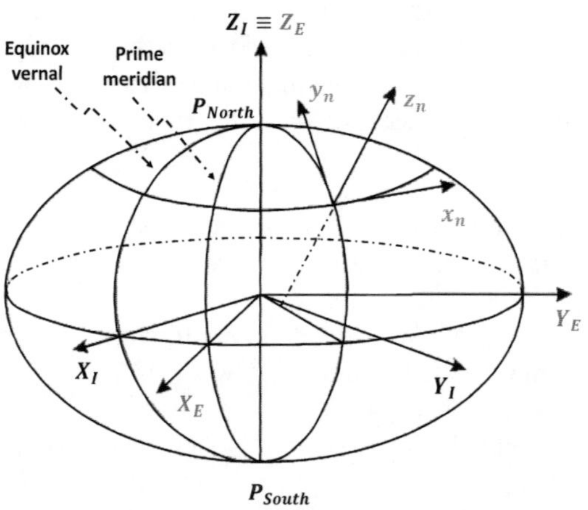

Figure 1. ECEF, ECI and ENU coordinate systems.

The equations that transform the position and velocity of the rocket from the rotating to the inertial coordinate system are, respectively,

$$\begin{bmatrix} P_X \\ P_Y \\ P_Z \end{bmatrix}_{ECI} = \begin{bmatrix} cos(\omega t) & -sin(\omega t) & 0 \\ sin(\omega t) & cos(\omega t) & 0 \\ 0 & 0 & 1 \end{bmatrix} \begin{bmatrix} P_X \\ P_Y \\ P_Z \end{bmatrix}_{ECEF} \quad (3)$$

$$\begin{bmatrix} V_X \\ V_Y \\ V_Z \end{bmatrix}_{ECI} = \begin{bmatrix} cos(\omega t) & -sin(\omega t) & 0 \\ sin(\omega t) & cos(\omega t) & 0 \\ 0 & 0 & 1 \end{bmatrix} \begin{bmatrix} V_X \\ V_Y \\ V_Z \end{bmatrix}_{ECEF} + \begin{bmatrix} -\omega sin(\omega t) & -\omega cos(\omega t) & 0 \\ \omega cos(\omega t) & -\omega sin(\omega t) & 0 \\ 0 & 0 & 0 \end{bmatrix} \begin{bmatrix} P_X \\ P_Y \\ P_Z \end{bmatrix}_{ECEF} \quad (4)$$

where $\omega = 7{,}2921159\ x\ 10^{-5}$ rad/s is the angular velocity of the Earth and t the flight time of the rocket.

3.1. Flight restrictions and simplifying assumptions
The constraints imposed in the algorithm that calculates the trajectory are as follows:

- the second stage is ignited immediately after the end of the firing of the first stage
- the mass of the fairing is subtracted at 105 km of altitude
- the third stage is limited to offering only horizontal velocity to the payload, i. e., the yaw correction maneuver ("dog leg") is not performed
- the azimuth angle is maintained above 89° throughout the flight
- drag coefficient is constant and equal to 0.5
- the influence of the wind is not considered
- eccentricity of the injection orbit is maintained below 10^{-4}

4. Results
The results obtained for launching payloads from 150 to 225 kg mass, in the eastern direction of CLA, are shown in figures 2 to 5. According to Equation 1, the larger the mass of payload carries, the lower is the acceleration delivered by the rocket. This causes the rocket to reach lower altitudes which, consequently, reduces the radius of the injection orbit. However, the smaller the radius of the orbit, the larger the velocity required to keep the microsatellite in orbit, as shown in Figure 2. A linear ratio are inverse to the payload increase and the injection altitude, but the increment in payload mass requires larger injection velocity.

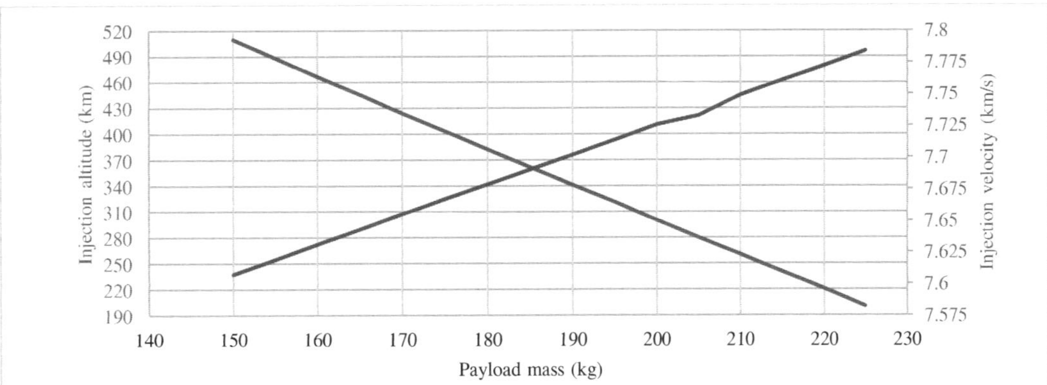

Figure 2. Altitude and velocity injection vs. payload mass.

The burning time of the three propellant stages of the VLM are well defined in Table 1. The duration of the ballistic flight phase is a parameter to be determined according to the restrictions imposed for the circular orbit. At this stage of flight there are no thrusters acting in the vehicle, it is a free or ballistic flight. The high altitudes and lower density atmosphere reduces the drag effect and the gravity is the main force acting along the trajectory. The increasing in the payload mass reduces the horizontal range of the final stage rocket, as well as the free flight time, because the mass increment generates a faster gravity turn, so reducing the horizontal velocity (Figure 3).

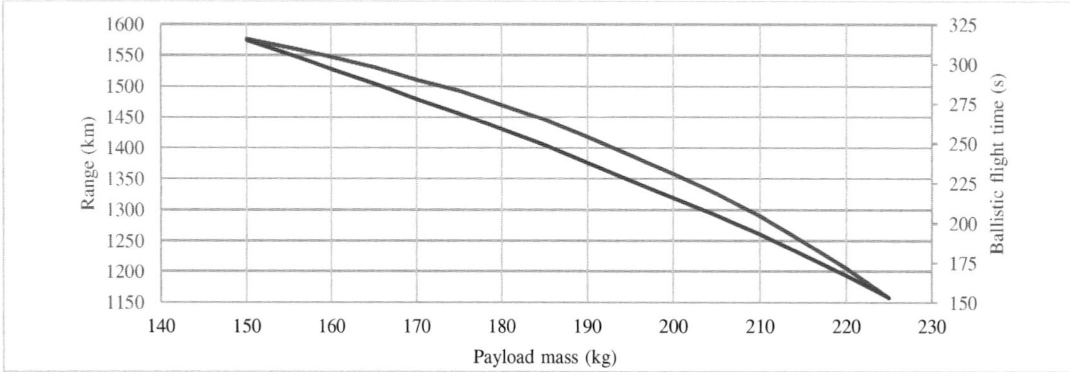

Figure 3. Range and ballistic flight duration vs. payload mass.

The geographic coordinates which correspond to the position where the payload separation occurred is shown in Figure 4. The small variation in latitude with respect to the position of the CLA is due to the constraint imposed on the azimuth angle of the VLM ($\beta = 90°$).

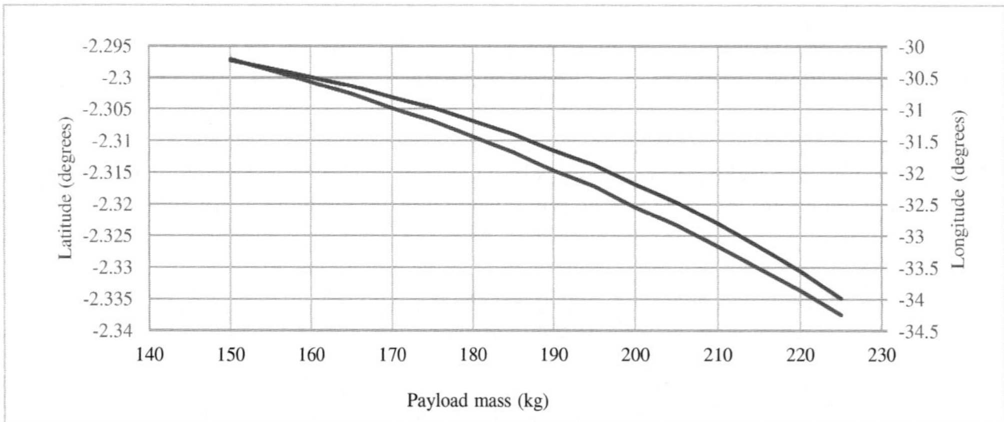

Figure 4. Latitude and longitude vs. payload mass at the moment of injection into orbit.

The second stage burnout altitude and FPA reduces due to the increment in the payload mass (Figure 5). These variations generate changes in the trajectory and in the final orbital injection.

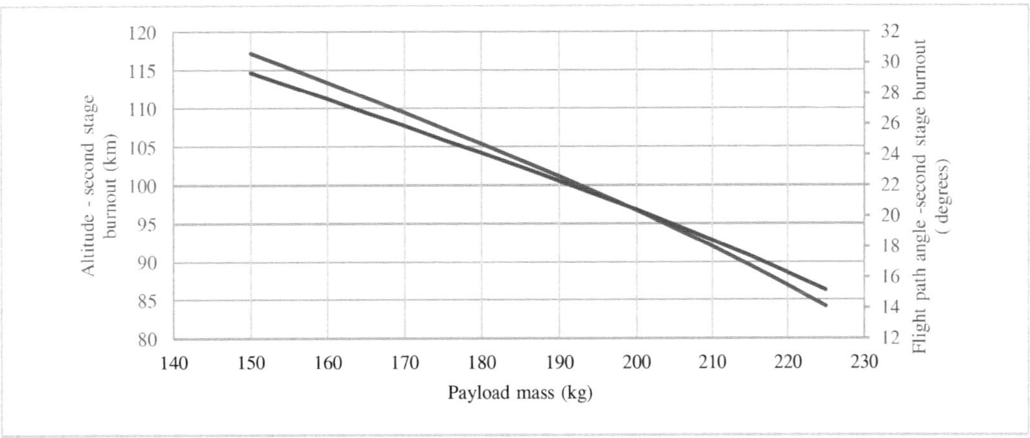

Figure 5. Altitude and flight path angle vs. payload mass at second stage burnout.

To determine the final ballistic time and the payload injection, the selected parameter was the FPA. A value around 0° is critical for the third stage ignition to guaranty the impulsive force in the same direction of the velocity and the increase of the final velocity. This relation is quasi-linear and presented in Figure 6.

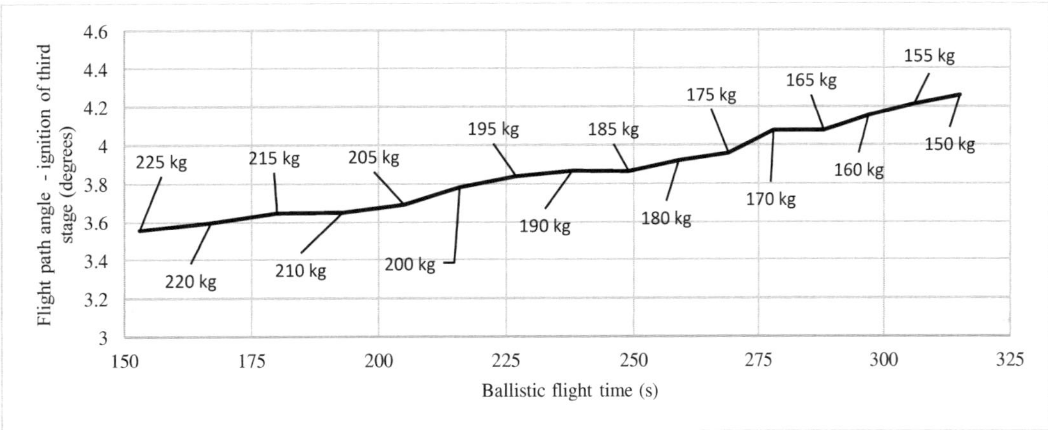

Figure 6. Flight path angle at ignition of third stage and payload mass vs. ballistic flight duration.

5. Conclusions and recommendations

The implementation of the mathematical model and the VLM performance in a computational code to simulate the flight trajectory are useful to determine the payload orbital conditions and the variations in the trajectory caused by the possible payload masses. In this case, the results show an increment in the orbital velocity proportional to the payload increment and a reduction in the injection altitude. The horizontal range and the ballistic time are inversely proportional to the payload mass increment. The payload mass increment generates altitude losses in the second stage because the gravity attraction and the lower acceleration generates a higher FPA changes due to the gravity turn.

The small satellite industry requires a new configuration in the Launch Vehicles. The paper presents in 3DOF the trajectory and the orbital injection changes due to the small payload variations. The Brazilian VLM was selected for the study. The simulated data are useful to compare the futures flight data in the VLM´s missions. Is recommended an implementation of the mathematical model in 6DOF for better approximations.

Acknowledgements
The authors wish to express their appreciation for the support provided by the National Council for the Improvement of Higher Education (CAPES), the National Institute for Space Research (INPE), the grants # 406841/2016-0 and 301338/2016-7 from the National Council for Scientific and Technological Development (CNPq), and the grants # 2011/08171-3, 2016/14665-2, from Sao Paulo Research Foundation (FAPESP).

References
[1] AEB, National Program of Space Activities PNAE: 2012-2021, Brasilia, 2012.
[2] Souza C H M *Development of a computational tool for use in the design of launch vehicles and application to the Vehicle Microssatellite Launcher*. 65p. Undergraduate thesis – Aeronautics Institute of Technology (ITA), São José dos Campos – Brazil, 2012.
[3] Silveira G *Development of a computational tool for flight simulation of launch vehicles*. 151p. Master Thesis – National Institute for Space Research (INPE), São José dos Campos – Brazil, 2014.
[4] Mota F A S *Modeling and simulation of launch vehicles using object-oriented programming.* 181p. Doctoral thesis – National Institute for Space Research (INPE), São José dos Campos – Brazil, 2015.
[5] Souza C H M *Analysis of the VLM-1 first stage separation dynamics*. Master Thesis – National Institute for Space Research (INPE), São José dos Campos – Brazil, 2015.
[6] Miranda D J F *Trajectory optimization of a satellite launch vehicle using pseudo-spectral methods*. 170p. Undergraduate thesis – Aeronautics Institute of Technology (ITA), São José dos Campos – Brazil, 2012.
[7] NOAA; NASA; US Air Force. US Standard Atmosphere, 1976. Washington, D. C.: Government Printing Office, 1976.
[8] Mooij, I. E. The motion of a vehicle in a planetary atmosphere, Delft University of Technology, Faculty of Aerospace Engineering, Delft – Netherlands, pages 29 - 31, 1994.
[9] Fehlberg, E. Classical Fifth-, Sixth-, Seventh-, and Eighth-Order Runge Kutta Formulas with Stepsize Control – NASA Technical Report-287. George C. Marshall Space Flight Center, Huntsville, Alabama – USA, 1968.
[10] Benavoli A, Chisci L, and Farina A 2007 Tracking of a Ballistic Missile with a-priori information *IEEE Transactions on Aerospace and Eletronic Systems* **vol** 43 issue 3 pp. 1000-1016. DOI: 10.1109/TAES.2007.4383589

IOP Conf. Series: Journal of Physics: Conf. Series **911** (2017) 012013 doi:10.1088/1742-6596/911/1/012013

Relative Positioning Evaluation of a Tetrahedral Flight Formation's Satellites

W F C Mahler [1] **E M Rocco** [1] **D P S Santos** [2]

[1] Space Mechanics and Control Division, National Institute for Space Research, São José dos Campos, Brazil
[2] São Paulo State University (Unesp), São João da Boa Vista, Brazil

wagner.mahler@inpe.br

Abstract. This paper presents a study about the tetrahedral layout of four satellites in a way that every half-orbital period this set groups together while flying in formation. The formation is calculated analyzing the problem from a geometrical perspective and disposed by precisely adjusting the orbital parameters of each satellite. The dynamic modelling considers the orbital motion equations. The results are analyzed, compared and discussed. A detection algorithm is used as flag to signal the regular tetrahedron's exact moments of occurrence. To do so, the volume calculated during the simulation is compared to the real volume, based on the initial conditions of the exact moment of formation and respecting a tolerance. This tolerance value is stablished arbitrarily depending on the mission and the formation's geometrical parameters. The simulations will run on a computational environment.

1. Introduction

Satellite formation flight is a set of spacecrafts acting cooperatively in a mission, for example NASA's mission MMS [1]. However, to have cluster of satellites in formation flight is a precise and complex task, due to the number of factors that affect the orbital movement, such as disturbances, attitude, the satellites' individual position etc.

This paper proposes the geometrical position of a four-satellite formation flight and the initial conditions needed to attain this arrangement. The satellites are positioned in a tetrahedral geometry formation in a way that it re-forms periodically without the need for orbital maneuvers or control action acting upon the movement prioritizing such formation. This arrangement occurs around the Earth and the dynamics of movement is relative to a geocentric frame of reference. The orbital elements of the four satellites are calculated and provided to the simulation's computational environment. This computational environment is known as Spacecraft Trajectory Simulator (STRS) [2] [3] [4] [5] [6], it will be responsible for the orbital motion dynamics simulation. The body's relative position during the formation of a regular polyhedron is analyzed by a detector proposed in order to indicate every regular tetrahedral geometry occurrence, within the tolerances admitted in an arbitrary manner and which vary according to the mission's needs. This work is illustrated in Figure 1.

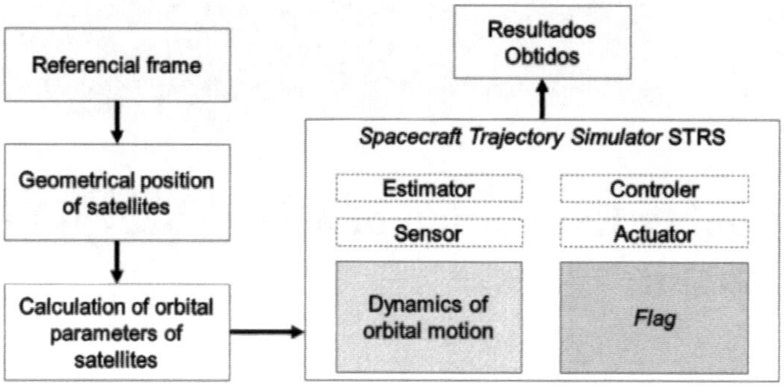

Figure 1. Conception of the stages attributed during this work's realization.

2. Mathematical Modelling

To form the tetrahedron's base the satellites S_1, S_2, e S_3 are considered to have circular orbits around the Earth. For the tetrahedron to be regular, the orbit of the satellite S_4 has to be elliptical [7]. The orbital motion is described in a geocentric frame of reference, whereas the satellites' relative movement is analyzed through reference frames centered on the satellites. Figure 2 presents a coordinate system with inertially defined axes positioned on the center of the Earth and a relative coordinate system positioned on the center of mass of all four satellites. The satellites S_1 and S_2 are in the same orbit and S_3 has inclination (i_3) from the orbital plane of S_1 and S_2. The forth satellite's inclination from the orbital plane is $i_4 = {}^{i_3}/_3$ and because it's an elliptical orbit, the orbit of satellite S_4 has low eccentricity, but the same orbital period as the other satellites, that is, $T_1 = T_2 = T_3 = T_4$.

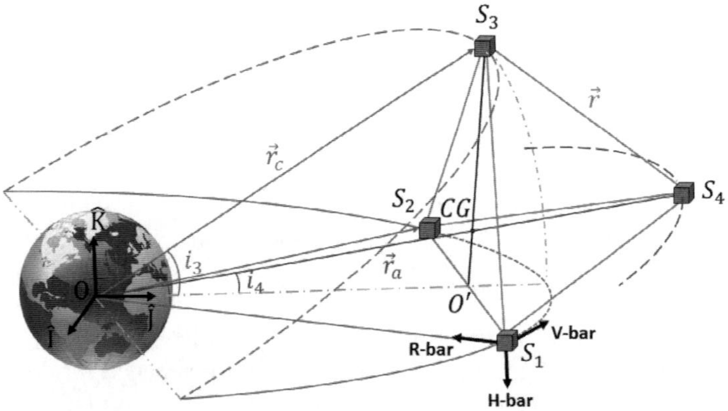

Figure 2. Tetrahedral configuration at the apogee of the S_4 orbit.

To determinate the body's mean anomalies, S_1 e S_2 were defined as being on the same circular orbit and the angle between them is the mean anomaly's variation, according to the equation.

$$\Delta M = \left(\frac{\mu}{r_c^3}\right)^{\frac{1}{2}} (t_2 - t_1) \tag{1}$$

Where t_1 and t_2 are the times relatives to S_1 and S_2 and μ the standard gravitational parameter. The satellite S_3 locates in the midpoint of the distance between S_1 and S_2 and its mean anomaly is calculated as following:

$$M_3 = \frac{\Delta M}{2} \tag{2}$$

The orbit of S_4 is elliptical and it was defined that the tetrahedron would form at the apogee, that is, $M_4 = 180°$.

The tetrahedron volume varies according to de satellites' motion. However, this volume is known as the regular tetrahedron forms. During the entire simulation, the calculated volume is compared to the nominal volume V_n of a regular tetrahedron. When both volumes are equal, the flag signals 1 showing that the tetrahedron has formed within an admissible tolerance. The algorithm responsible for this flag is shown on Figure 3.

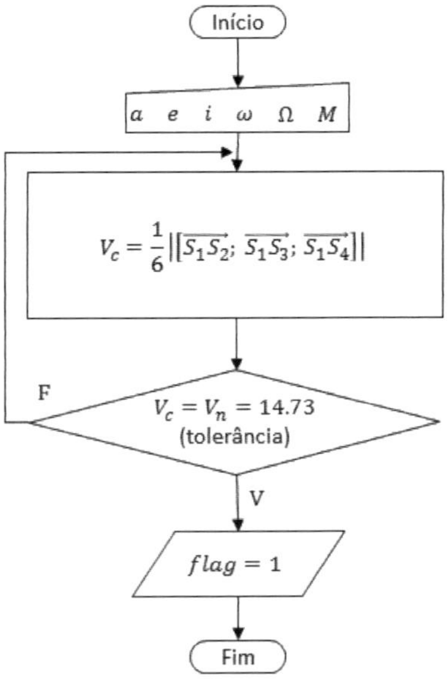

Figure 3. Regular tetrahedron formation detection algorithm based on volume.

3. Results

For the satellites in tetrahedral formation flight orbital motion simulation, the perturbative effects were not considered a priori, because the relevant fact is to evaluate the satellites' geometrical arrangement during flight. Simulation data is presented in Table 1.

Table 1. First simulation input parameters.

Simulation Data – Simulation I				
	S_1	S_2	S_3	S_4
h (m)	800000	800000	800000	800000
r (m)	5000	5000	5000	5000
a (m)	7178100	7178100	7178100	7178100
e	0	0	0	0.0005687383463373053
i (°)	0	0	0.034562999271458	0.011520999757153
Ω (°)	0	0	0	0
ω (°)	0	0	0	270
M (°)	0	0.0399099558281	0.01995497791405	90.01995497791405
Simulation time = 6055 s (one nominal orbital period)				
Tetrahedron's volume = 14.73 km^3				
Maximum error allowed relative to the nominal edge (5000 m) = 50 m				
Maximum error allowed relative to the diferences between positions = 100 m				
Maximum error allowed relative to the nominal volume = 0.05 km^3				

The apparent motion of the satellites is observable from the reference frame on the center of S_1. It is noted from that reference frame the movement of the other satellites, while S_1 remains stationary. It is observable that S_2 finds itself stationary relative to S_1, while S_3 and S_4 are not. That happens because S_2 is on the same orbit as S_1, while S_3 and S_4 are not and, besides, both possess orbital inclination relative to S_1 orbital plane. The period of the four bodies' orbits is the same, demonstrating for a fact that there is occurrence of a tetrahedron (Figure 4).

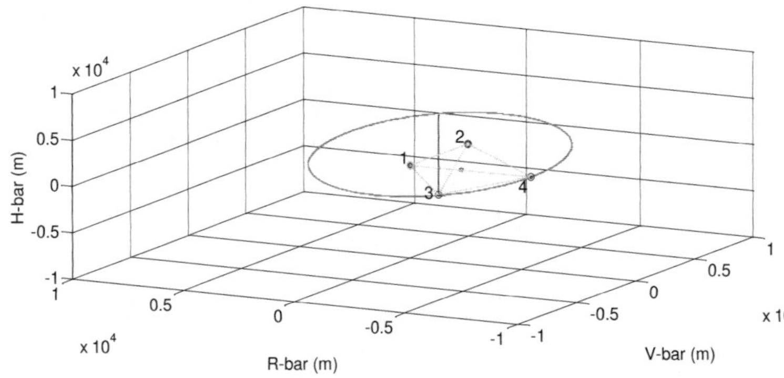

Figure 4. Tetrahedral flight formation's satellites observed from de coordinate system centered on S_1 and the trajectories assumed by S_3 and S_4 relative to S_1.

The detection flag (Figure 5) shows that the regular tetrahedron happens twice during orbital motion.

IOP Conf. Series: Journal of Physics: Conf. Series **911** (2017) 012013 doi:10.1088/1742-6596/911/1/012013

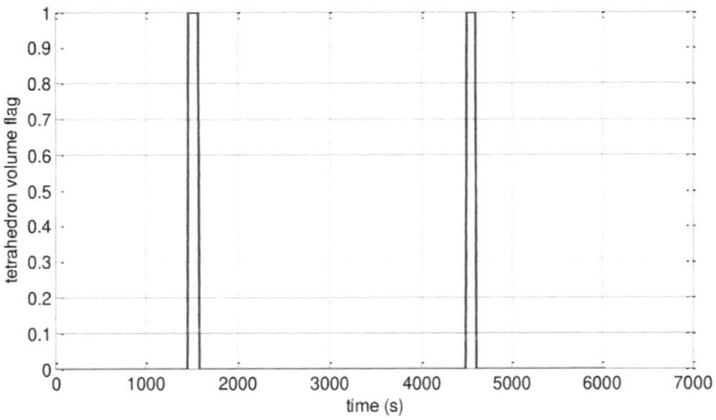

Figure 5. Two moment regular tetrahedron occurrence signalization flag.

The regular tetrahedron formation's periodicity during the orbital motion can be seen in Figure 6.

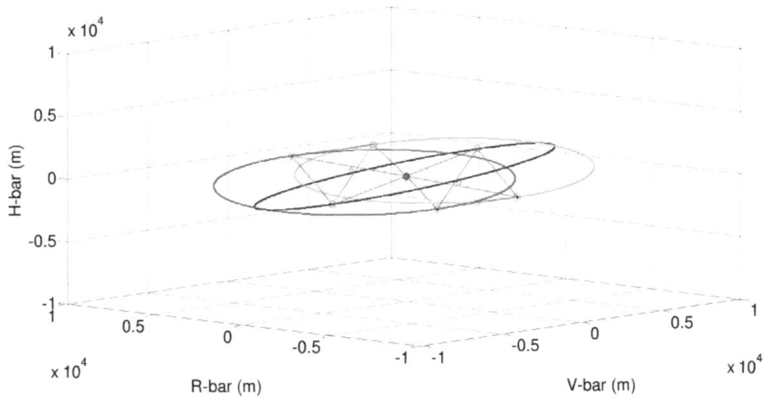

Figure 6. Regular tetrahedron formation's periodicity.

The movement of satellite S_3 is of deviation and elevation from S_1 orbital plane and at 1500 s, the segment $\overline{S_1S_3} = 5000$ m and the triangle formed by $S_1S_2S_3$ is formed. A few moments after this conjunction of satellites, their relative position degrades and the bodies reach that configuration again at 4500 s (Figure 7).

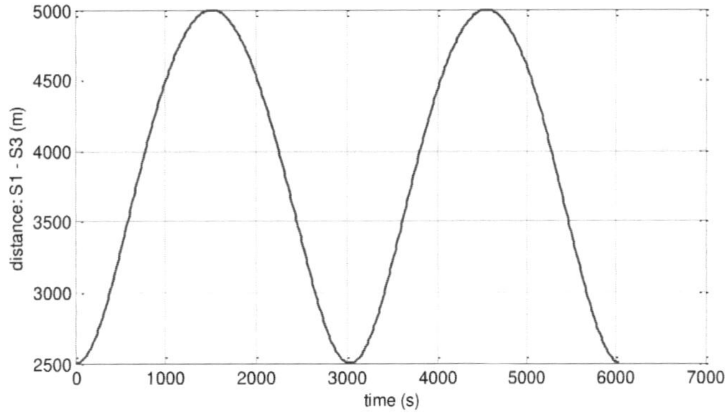

Figure 7. S_3 position's variation relative to S_1.

The composition of the ascension movement and the transversal velocity of S_3 satellite's orbit, when observed from the reference frame of S_1 generates a Lissajous figure (Figure 8), due to the apparent motion.

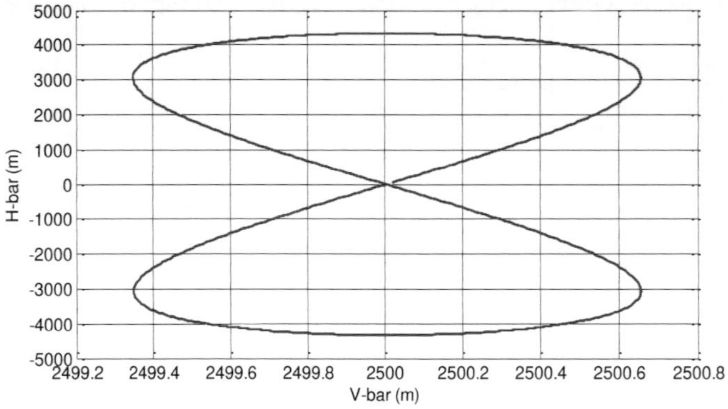

Figure 8. S_3 position's variation relative to S_1, from a local perspective.

4. Conclusion

This paper enables an analysis of the relative behavior between satellites in a geometrical tetrahedral flight formation's satellites. The performed simulation shows that the precise positioning of the satellites is relevant for the tetrahedron's formation and its periodical maintenance. The simulation demonstrates that the satellites group together at each half period of orbit, first on the apogee then on the perigee. Even though it is most of the time in a tetrahedral configuration during orbital motion, the regular tetrahedron only occurs twice. It is worth noting that the results obtained consider ideal situations, thus allowing the four bodies' orbital parameters calculation.

References

[1] Schmid D et al. (2016) A comparative study of dipolarization fronts at MMS and Cluster *Geoph. Research Letters* 43(12) 6012-6019 doi:10.1002/2016GL069520

[2] Rocco E M 2008 Perturbed orbital motion with a PID control system for the trajectory *XIV Col. Bras. de Dinâmica Orbital* (Águas de Lindóia: Brazil)

[3] Rocco E M 2008 Analisys of the Trajectory due to the Terrestrial Albedo Applied to some Scientific Missions *Int. Conf. on Mat. Problems in Engineering* (Genova: Italy)

[4] Rocco E M 2012 Controle de trajetória com propulsão contínua para missões do tipo drag-free *VII Cong. Nac. de Engenharia Mecância* (São Luís: Brazil)

[5] Rocco E M 2013 Automatic correction of orbital elements using continuous thrust controlled in closed loop *J. Phys.: Conf. Series* **465**

[6] Rocco E M 2015 Gravitational disturbances generated by the Sun, Phobos and Deimos in orbital maneuvers around Mars with automatic correction of the semi-major *J. Phys.: Conf. Series* **641**

[7] Mahler W F C, Rocco E M and Santos D P S 2016 Avaliação do Posicionamento Relativo dos Satélites de uma Formação de Voo Tetraédrica *XVIII Col. Bras. de Dinâmica Orbital.*

IOP Conf. Series: Journal of Physics: Conf. Series **911** (2017) 012014 doi:10.1088/1742-6596/911/1/012014

Study of orbital transfers with time constraint and fuel optimization

L G Meireles and E M Rocco

National Institute for Space Research, INPE, São José dos Campos, Brazil

meireleslg@gmail.com; evandro.rocco@inpe.br

Abstract. Over the decades, with the advance in computational power and the development of new optimization techniques, a new opportunity arose for a reviewed approach and analysis of already existing solutions to orbital transfer problems. With that in mind, a study was proposed over Lambert's problem of bi-impulsive orbital transfers with time constraint for the execution of minimum fuel maneuvers between coplanar and non coplanar elliptical orbits. In order to achieve this goal, a genetic search algorithm was developed to determine the minimum fuel trajectory with an iterative variation of two Keplerian elements of the final orbit, one of them being the mean anomaly and the other a choice between the eccentricity, the inclination or the argument of periapsis. A number of six simulations were performed under these conditions, using different values - on each simulation - for the Keplerian elements of the final orbits not being iteratively varied. Remarkable sections were noted regarding the behaviour of fuel consumption values of the minimum fuel trajectories. One of which was a noted relation between the lowest cost minimum fuel transfers and the ratio of the semi-major axis and eccentricity of these transfers' initial and final orbits, considering both the orbits have the same value of argument of periapsis. Amongst all simulations, it was important to identify the choice between short or long way for the determination of the minimum fuel transfer, analysing the implications of that choice on the transfer time.

1. Lambert's Problem

The objective of this paper is the presentation of a study on methods of bi-impulsive orbital transfers with time constraint for the realization of fuel optimization between coplanar and non coplanar elliptical orbits. As an approach to calculate the transfer orbital maneuvers was considered the same methodology used in [1-5], where the equations were taken from the solution of the Lambert's problem presented in [6-8]. This method allows to determine the transfer orbit, with time constraint, that connects the terminal points within the initial and final orbits. This is known as the Two Point Boundary Value Problem (TPBVP) as shown in a simplified way in figure 1.

Initially, after defining the initial and final position vectors r_0 and r_1, it is necessary to determine their norm. With them, it is possible to determine the transfer angle for the short way, in equation 1, or the long way, in equation 2.

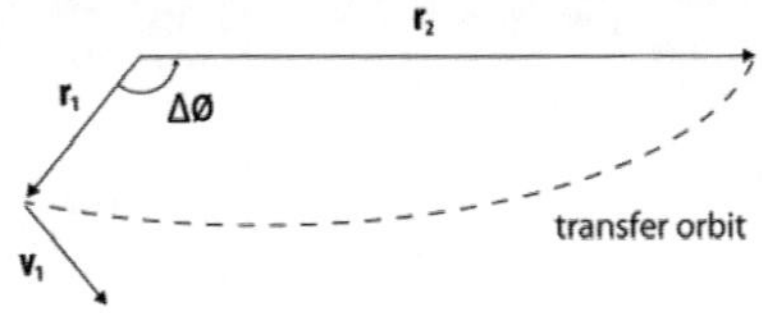

Figure 1. Two point boundary value problem

$$\Delta\Phi = cos^{-1}\left(\frac{\vec{r_0}\cdot\vec{r_1}}{r_0 r_1}\right) \tag{1}$$

$$\Delta\Phi = 2\pi - cos^{-1}\left(\frac{\vec{r_0}\cdot\vec{r_1}}{r_0 r_1}\right) \tag{2}$$

With the transfer angle, determine the *A* constant value as seen in equation 3.

$$A = \sqrt{r_0 r_1\left(1 + cos(\Delta\Phi)\right)} \tag{3}$$

By iteratively estimating a *z* constant value inside the interval $[0,4\pi^2]$, it is possible to determine a *C* and *S*, alongside a *y* and *x* constant, and calculate the estimated time, from equations 4 to 6.

$$C = \frac{1-cos(\sqrt{z})}{z} \qquad S = \frac{\sqrt{z}-sin(\sqrt{z})}{\sqrt{z^3}} \tag{4}$$

$$y = r_0 + r_1 - A\left[\frac{1-zS}{\sqrt{C}}\right] \qquad x = \sqrt{\frac{y}{C}} \tag{5}$$

$$\Delta t = \frac{1}{\sqrt{\mu}}\left[x^3 S + A\sqrt{y}\right] \tag{6}$$

After the difference between the estimated time and the determined time fit inside a tolerance value, a set of constants *f*, *g* and$\dot{g}$can be determined from use of equations 7 to 9.

$$f = 1 - \frac{y}{r_0} \tag{7}$$

$$g = A\sqrt{\frac{y}{\mu}} \tag{8}$$

$$\dot{g} = 1 - \frac{y}{r_1} \tag{9}$$

With these values, it is possible to calculate the velocity components on the initial and final points, with the use of equations 10, and the sub consequent velocity increment on those points, from equations 11.

$$\dot{\vec{r_A}} = \frac{1}{g}\left[\vec{r_1} - f\vec{r_0}\right] \qquad \dot{\vec{r_B}} = \frac{1}{g}\left[\dot{g}\vec{r_1} - \vec{r_0}\right] \tag{10}$$

$$\Delta v_1 = \dot{r_A} - \dot{r_0} \qquad v_2 = \dot{r_1} - \dot{r_B} \tag{11}$$

2. Search parameters

This study starts from the analysis of multiple solutions to the Two Point Boundary Value Problem (TPBVP), by variating the given transfer time for given terminal points, all of them serving as inputs of the TPBVP. The output of the TPBVP is the velocity increment necessary to achieve such a transfer. To obtain the TPBVP solution, Holland's canonical genetic algorithm [9-10] was implemented to find the z variable value, always contained in the $[0,4\pi^2]$ for elliptical problems. With this z variable value the velocity increment is determined.

It was evident from the results obtained from the iterative sweep of initial conditions for the TPBVP, that for each transfer time sweep there was a minimum value for the velocity increment. Therefore the genetic search algorithm was modified to search not for velocity increment of the given value of transfer time, but to search for the minimum velocity increment. Consequently, the transfer time is no more an input of the problem, it is an output. This approach made it easier to obtain the results for a minimum velocity increment. The graphical presentation of the results of these simulations, using the iterative sweep of the initial conditions for the TPBVP, facilitates their analysis.

The canonical genetic algorithm was implemented with a population of 50 individuals, which values were of a random variable z, and a random arithmetic recombination within the 25 strongest individuals over 50 generations with a 15% chance of random mutation. This procedure was repeated 8 times, thus a total of 8 universes with randomly distinct initial populations inside the $[0,4\pi^2]$ interval were considered, taking the strongest individual within all the generations and universes iterated. With this approach, it was possible to analyze the variation of two Keplerian elements in each simulation. One of which was the mean anomaly of the final orbit, in order to sweep all possible final positions, whilst shifting a second Keplerian element in order to understand how these changes affected the behavior of the minimum fuel transfers. It is worthwhile to note that the algorithm calculates the minimum fuel consumption for the short and long way trajectory and chooses the actual minimum fuel trajectory, indicating the long way choices as blue dots in the graphs.

The initial and final orbit's Keplerian elements are presented by table 1 and table 2, respectively. Table 2 indicates the range of variation of the Keplerian elements being varied in each simulation.

Table *1*. Initial orbits' Keplerian elements

a (R_{Earth})	e	i (°)	Ω (°)	ω (°)	M (°)
1.5	0.0000001	0	0	0	0

Table 2. Final orbits' Keplerian elements

	a (R_{Earth})	e	i (°)	Ω (°)	ω (°)	M (°)
Sim. 1	2	0.0000001 – 0.9	0	0	0	0 – 360
Sim. 2	2	0.3	0	0	0 – 360	0 – 360
Sim. 3	2	0.0000001	0 – 90	0	0	0 – 360
Sim. 4	2	0.0000001	0 – 90	45	0	0 – 360
Sim. 5	2	0,8	0	0	0 – 360	0 – 360
Sim. 6	5	0,3	0	0	0 – 360	0 – 360

3. Simulation results

The results of all the simulations can be seen in figure 2. Simulation 1 consists of 1000 points, simulations 2, 5 and 6 show 1200 results and simulations 3 and 4 determined 420 minimum fuel trajectories. Each point of the data plots displayed in figure 2 presents a minimum velocity increment for a transfer time determined by the genetic algorithm search. The multiple points are a consequence of the variation of the terminal points, which in turn are a function of the keplerian elements of the final orbit. For each sweep of keplerian elements, the minimum velocity increment found by the genetic algorithm presents a pattern of variation, indicated by the plots in figure 2. The analysis of theses plots may indicate regions of lower values of minimum velocity increment for different terminal points.

By knowing that all the keplerian elements, except for the semi-major axis of orbit, have a finite well determined interval of possible values, the sweep was determined in order to cover all of the values the keplerian element being varied could achieve. Therefore, by covering all the values that could be physically achieved, the lower value regions for minimum velocity increment were identified as the minimum values for the interval of terminal points analyzed.

In summary, the genetic algorithm determined the value of the minimum velocity increment for an already given terminal point. These terminal points were then iteratively varied, by sweeping the final orbit's keplerian elements. This process is proven useful in mission definition where the final orbit is open to determination. By identifying the lower value regions of minimum velocity increment, the final orbit's keplerian elements can be determined more easily. Nevertheless, it is imperative to say that mission determination is a multi objective search, where a minimum velocity increment is desirable but not the only target to be achieved. A multiobjective approach that could be applied in this case, to the search for a compromise solution that considers the simultaneous optimization of the fuel consumption and the orbital transfer time, can be found in [11-13]. However, the application of a multiobjective approach is beyond the scope of this work.

Additionally, careful considerations can be made for each of the simulation results. For example, if analyzed attentively, simulation 4 results in figure 1 shows two high fuel consumption boundary regions between the choices of short and long way. Figure 3 illustrates a remarkable trajectory close to these boundary regions. It is noticeable that a difference in the right ascension of the ascending node of the initial and final orbits raises conditions in which the final orbit can only be achieved, for transfer angles near 180°, by transfer orbits with very high inclination. This condition creates the highest Δv values of the simulation, explaining the source of the verified transition regions in figure 1.

4. Conclusion

A direct search for the minimum fuel consumption using Lambert's approach to the Two Point Boundary Value Problem, whilst calculating the transfer time relative to the minimum fuel trajectory determined, eases human analysis of results for simulations varying two Keplerian elements simultaneously. Considering this type of simulation, it is important to note the choice made by the search algorithm between the short and long way. It must be able to analyze both options in order to be able to choose the actual minimum fuel trajectory, instead of imposing a reversion in the direction of the maneuver to achieve the short way and calculate a considerably larger fuel consumption.

The variation of the final orbit's mean anomaly and consequent sweep of multiple final points inside this orbit, with the variation of other of its Keplerian elements and change on its shape and orientation, leads to surprising results. For example, a variation of the final orbit's eccentricity value, with a constant semi-major axis ratio between the initial and final orbits, added to equal values of argument of periapsis between them and quasi-circular initial orbit, leads to a lower value region of minimum fuel consumption trajectories for a specific eccentricity value. This region depends on the initial and final orbits' semi-major axis ratio. Another example is the change in the right ascension of the ascending node of the final orbit for non coplanar orbits leads to higher value regions of minimum fuel consumption trajectories for maneuvers with angles of approximately 180°. This is the opposite effect, as shown by the Hohmann transfer, displayed by initial and final orbits with equal right ascension of the ascending node.

The analysis must be extremely cautious in regions of angles near 180° and 0° or 360°. As the angles approach these values, numerical errors are far greater, making it advisable to discard a group of results that fell under these conditions. Nevertheless, the inclusion of these regions in the search parameters is interesting in order to help understand the global behavior of the simulation when approaching these extremes values.

A broad variety of possible simulations is still open to further studies. A total of six simulations were implemented. Amongst these, three of them were codependent simulations with the variation of the same pair of Keplerian elements but with a change in a third element of the final orbit, providing independent results with different standards of behavior. A proposed new change of these values can be made to identify new standards of behavior or reinforce the ones already identified, or the variation of other pairs of Keplerian elements, keeping a similarity to other simulations. Such analysis requires a great computational power, a great amount of time and the development of a faster and more efficient algorithm, perhaps written in C language, but will unveil a broader understanding over Lambert's problem and all its possible approaches.

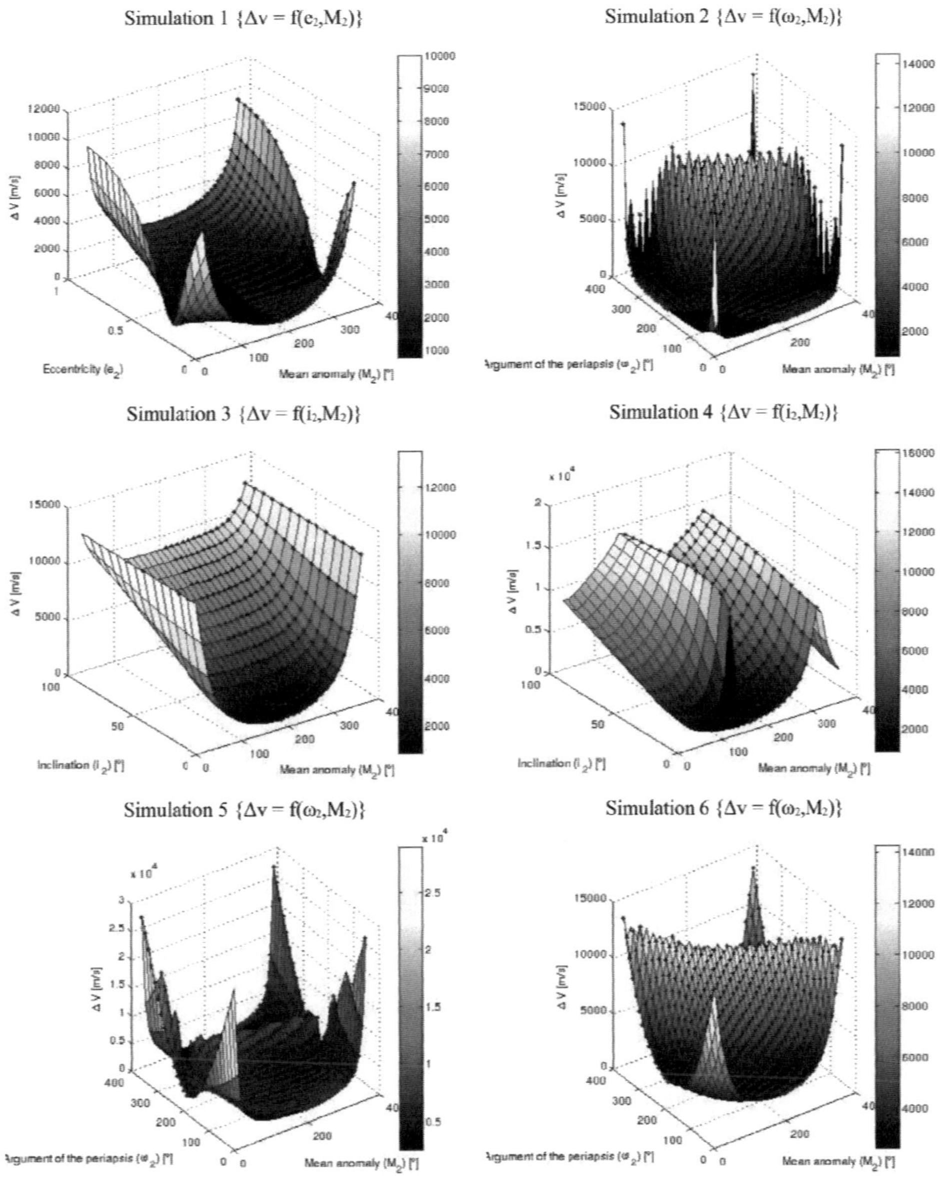

Figure 2. Graphical results of the Δv values of the minimum fuel transfers in function of e_2, i_2, ω_2 and M_2.

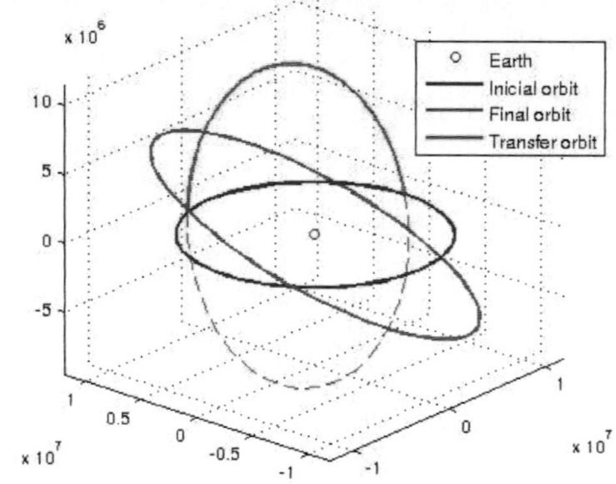

Figure 3. Remarkable trajectory of Simulation 4

References

[1] Rocco E M 2013 *Simulação de Manobras Orbitais Impulsivas com Empuxo Limitado e Controle de Trajetória em Malha Fechada* DINCON 2013 – Conferência Brasileira de Dinâmica Controle e Aplicações 13 to 17, October, 2013 Fortaleza - CE.

[2] Rocco E M 2015 *Gravitational disturbances generated by the Sun, Phobos and Deimos in orbital maneuvers around Mars with automatic correction of the semi-major axis.* Journal of Physics. Conference Series (Print) v. 641, p. 012027. doi: 10.1088/1742-6596/641/1/012027

[3] Arantes Jr G; Rocco E M; da Fonseca I M; Theil S 2010 *Far and proximity maneuvers of a constellation of service satellites and autonomous pose estimation of costumer satellite using machine vision.* Acta Astronautica v. 66 p. 1493-1505. doi: 10.1016/ j.actaastro.2009.11.022

[4] Rocco E M 2014 *Simulation of the effects due to the gravitational disturbances generated by the Sun, Phobos and Deimos in orbital maneuvers around Mars.* In: XXXV Iberian Latin American Congress on Computational Methods in Engineering - CILAMCE 2014, Fortaleza - CE - Brazil. Proceedings of CILAMCE 2014.

[5] Mota M L; Rocco E M 2014 *Comparação entre as abordagens impulsivas e contínua do sistema propulsivo em manobras orbitais em torno de um corpo não-esférico.* In: XXXV Iberian Latin-American Congress on Computational Methods in Engineering - CILAMCE 2014, Fortaleza - CE - Brazil. Proceedings of CILAMCE 2014.

[6] Bate R R; Mueller D D; White J E 1971 *Fundamentals of Astrodynamics.* Dove Public. Inc.

[7] Bond V R; Allman M C 1996 *Modern Astrodynamics: fundamentals and perturbation methods* New Jersey: Princeton University Press.

[8] Battin R H 1999 *An Introduction to the Mathematics and Methods of Astrodynamics Rev ed.* Reston: AIAA Educational Series.

[9] Cunha A G; Takahashi R; Antunes C H 2012 Manual de Computação Evolutiva e Metaheurística. Coimbra: [s.n.] ISBN 978-989-26-0583-8

[10] Holland J 1992 *Adaptation in Natural and Artificial Systems.* Cambridge, MA: MIT Press.

[11] Rocco E M, Souza M L O, Prado A F B A. 2013 *Station Keeping of Constellations Using Multiobjective Strategies.* Mathematical Problems in Engineering doi:10.1155/2013/476451

[12] Rocco E M; Souza M L O; Prado A F B A 2003 *Multiobjective optimization applied to satellite constellations I: formulation of the smallest loss criterion.* In Proceedings of the 54st International Astronautical Congress (IAC'03), Bremen, Germany, September 2003.

[13] Rocco E M; Souza M L O; Prado A F B A 2005 *Multiobjective optimization applied to satellite constellations II: initial applications of the smallest loss criterion.* In 4th international workshop on satellite constellations and formation flying, pp. 14–16, National Institute for Space Research, São José dos Campos, Brazil, February 2005.

Orbital and attitude evolution of SCD-1 and SCD-2 Brazilian satellites

J O Murcia[1], V Carrara[1,2], and H K Kuga[1,2]

[1] National Institute for Space Research (INPE), Av. dos Astronautas 1.758, Jd. Granja, CEP 12227-010 São José dos Campos, SP, Brazil.
[2] Aeronautics Institute of Technology, ITA, São José dos Campos, SP, Brazil.

Correspondence author: helio.kuga@inpe.br

Abstract. The SCD-1 and SCD-2 satellites were launched in 1993 and 1998, respectively, with use of the Launcher "Pegasus" of the OSC (Orbital Sciences Corporation). 21 and 16 years later, the satellites are still in orbit around the Earth and providing data for users. Mission and Operational data from Satellite Tracking Center Network are stored in mission files in the Satellite Control Center (SCC) and made available to the users. The SCC also stores history files of the satellite orbit and attitude ephemeris, besides the on-board telemetry, temperatures, equipment status, etc. This work will present some analysis of the orbit ephemeris evolution based upon the Two-Line Elements sets (TLE´s) obtained from NORAD (North American Aerospace Defense Command). Attitude evolution along time is also presented for both satellites from SCC data. The orbit decay will be explained as resulting mainly due to the solar activity during the satellite lifetime. This work aims to report the history of more than 20 years of continuous operation of SCD-1 and SCD-2. At the end, an estimation of the orbital decay is forecast with the use of NASA's DAS software.

1. Introduction

On February 9th, 1993, the first Brazilian Environmental Data Collecting Satellite (SCD-1) was launched by the OSC launcher named "Pegasus" from an airplane. With 110 kg mass, it was placed into an orbit of 760 km altitude, circular, with inclination of 25°. In October 23rd of 1998, the SCD-2 was injected by "Pegasus" again in a similar orbit with altitude of 756 km, inclination of 25° and RAAN (Right Ascension of Ascending Node) of 220° (around 180° far from SCD-1 RAAN at the injection time) [1]. Both satellites were conceived to debut Brazil in space technology, under the Complete Brazilian Space Mission (MECB) program. The SCD satellites collect in real time environmental data from automated remote ground platforms. Collected data such as atmospheric properties, weather conditions, environment sensing, are broadcast to the mission ground station, located in Cuiaba, Brazil and disseminated to the interested community. Both satellites were deployed in a Low Earth Orbit (LEO) and were designed and manufactured by the National Institute for Space Research (INPE) in São José dos Campos, São Paulo, Brazil. More information related with the nominal value of the orbital elements are shown in Table 1. The SCD's are spin stabilized, with a passive nutation damper to reduce any residual nutation of the satellites. Attitude control of the spin axes is achieved by means of a spin plane magnetic air core coil, which is switched by on-ground commands. Additionally, SCD-2 has a spin axis magnetic coil switched by one on-board electronics to control its angular rate around 34 rpm. These satellites are still operating after more than 20 years and during this time orbit and attitude perturbations as well as attitude maneuvers affected their initial orbit and attitude. The large amount of orbit and attitude data provides a valuable source of information to study and to analyse the time evolution of orbit and attitude parameters.

The main objective of this work is to present the time history of the SCD satellites orbit and attitude, and to present some discussions about the evolution of these parameters.

Table 1. Nominal Orbital Elements at Injection times [1] [2].

Orbit Element	SCD-1	SCD-2
Semi-major axis (m)	7135839	7133898
Eccentricity	0.004568	0.000756
Inclination (°)	24.972	24.987
RAAN (°)	180	219.912
Argument of perigee (°)	347.75	348.543
Mean anomaly (°)	348.42	124.479

To analyse the variation in the orbital elements, two types of data were collected: the first are the TLE´s (Two-Line Elements set) from NORAD site (celestrak.com) and the second was the orbital history from Satellite Control Center, SCC, (at INPE, in São José do Campos, Brazil). The data from SCC was compared with the NORAD data to search for differences in orbital elements. Critical points where the values change rapidly in short time intervals were selected, and the perturbations that caused those changes are analysed in the next sections. The attitude of both satellites was obtained from SCC historical records and represents the satellites space orientation, angular speeds and the magnetic coils switching times. The data packages were obtained through the SCC, hourly, from February 10/1993 to October 24/2014, covering more than 21 years of data for SCD-1 satellite and 16 years for SCD-2. TLE packages were collected for the same period, containing daily orbital evolution.

2. Orbital elements evolution

The orbit of SCD-1 was specified to be circular, at 750 km altitude and with inclination of 25 degrees with respect to the Equator plane, which favours the mutual contact time of the ground remote platforms in Brazil, the satellite and the Cuiabá Ground Station. The SCD-2 orbit was chosen also to be circular, at 750 km altitude, 25 degrees inclination and with the right ascension of the ascending note (RAAN) 180 degrees apart from SCD-1 (at the time of launching), to increase the daily coverage of both satellites. The initial semi-major axis for SCD-1 just after launch was 7135.8 km. In 21 years the orbit height decreased by almost 12 km, due mainly to the atmospheric drag, as can be seen in figure 1. The orbit decay was increased between years 1999 to 2003, due to the solar maximum activity period, shown in figure 2. SCD-2 was injected at 7134.5 km semi-major axis and decreased some 10km.

It is known that the air density, which causes the decay of the semi-major axis, depends on the solar flux at 10.7 cm of wavelength, and this flux is higher at the solar maximum activity, that occurs in a 10.6 years interval [3] [4]. In 2012 a second solar maximum begun, so the altitude decay rate of both SCD-1 and 2 have increased again. During the maximum solar activity, the decay rate was around 2.3 km/year.

The eccentricity of both satellites is shown in figure 3. It can be noted that the orbit eccentricity presents a slight tendency to decrease toward zero, due also to the aerodynamic drag, that causes the apogee to decay faster than the perigee. The SCD-1 orbit eccentricity is very small and nearly circular, around 0.005, or 70 km of difference between the apogee and perigee heights. The eccentricity of SCD-2 orbit is around 0.002.

IOP Conf. Series: Journal of Physics: Conf. Series **911** (2017) 012015 doi:10.1088/1742-6596/911/1/012015

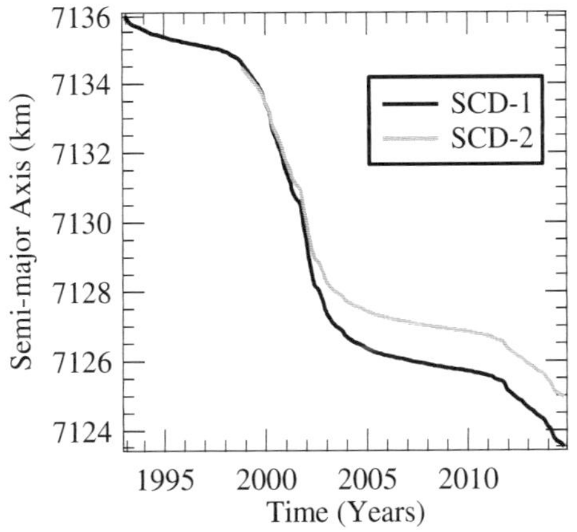

Figure 1. Semi-major Axis, NORAD data.

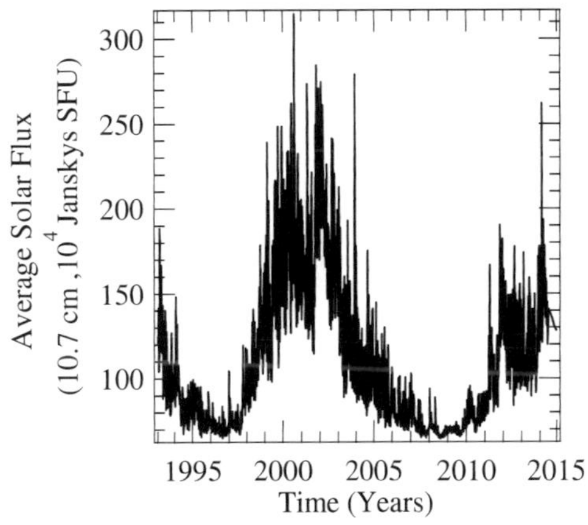

Figure 2. Solar Flux data.

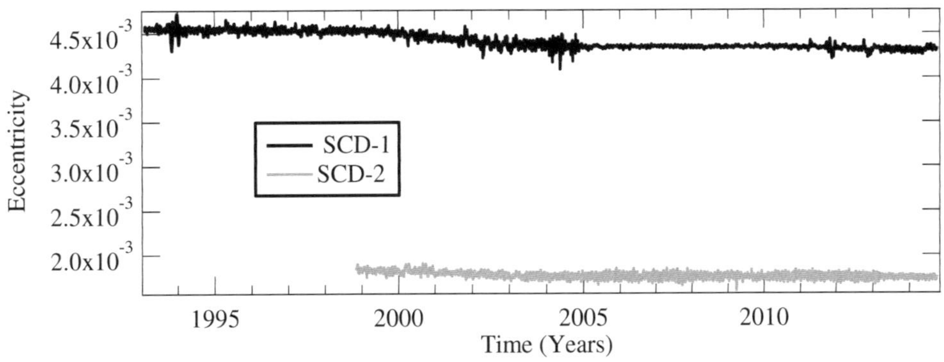

Figure 3. Satellites Eccentricity, NORAD data.

Due to injection errors, the orbit inclination of the satellites is also slightly different, around 24.970 degrees for SCD-1 and 24.996 degrees for SCD-2, as shown in figure 4. The major perturbations in the inclination are the Sun-Moon gravity forces and the Earth's non-uniform gravitational field. The inclination deviation of SCD-1 in 2004, as can be seen in figure 4, increased suddenly probably due to some sort of orbit resonance with the geo-gravitational field, but it has yet to be proved.

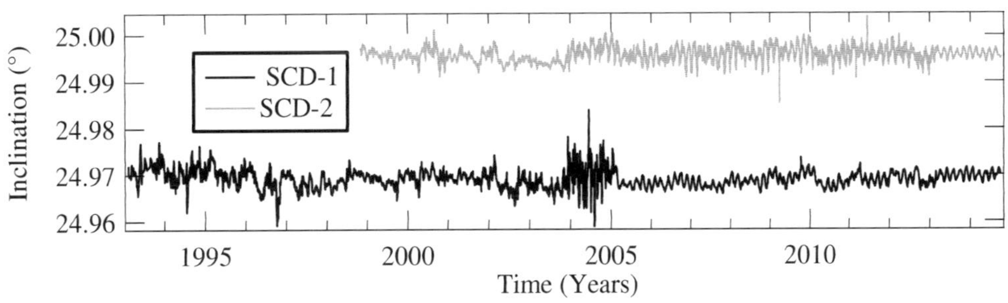

Figure 4. Satellites Inclination, NORAD data.

When the SCD-2 was placed in its orbit, its RAAN was close to 188° offset from the SCD-1 RAAN, with only 8° error from the specified value due to a small delay in the launch time. The non-symmetrical Earth's mass distribution due to the pole oblateness causes the orbit to precess around the Earth's rotation axis and so the RAAN changes with time. The rate of change in the right ascension node depends mainly of the orbit inclination and the semi-major axis, or the orbit height. As can be seen in figure 5, which shows the difference between the SCD-1 and SDC-2 right ascension angles, initially the difference remained constant about 180° since the SCD-2 was inject in orbit with almost the same height of SCD-1 (shown in figure 1). The decay of SCD-1 generates changes in the eccentricity of the orbit and the RAAN. Because of the decay of SCD-1, the difference between RAAN of both satellites has increased, reaching 38° amplitude from 180° at the beginning of 2015 (figure 5).

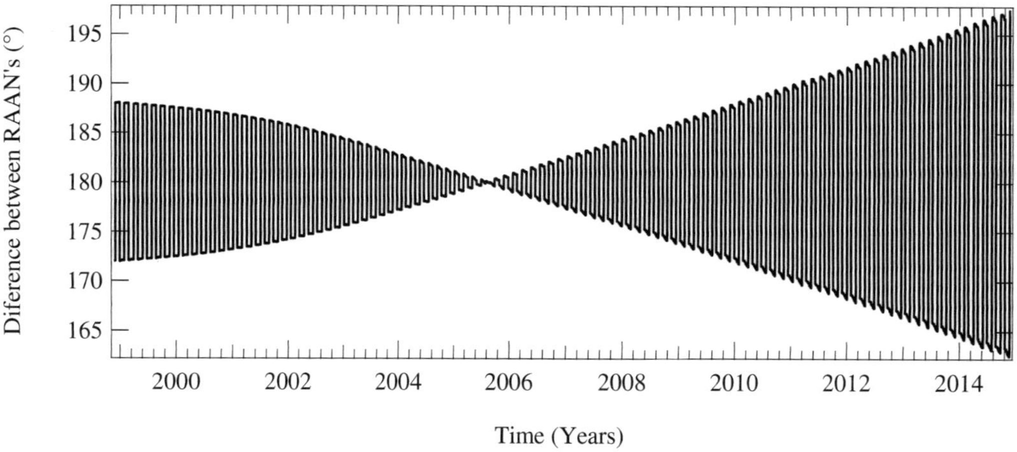

Figure 5. Difference between RAAN's, SCC data.

3. Attitude

The SCD-1 initial spin rate was around 120 rpm, and in less than two years of operations the rotation decayed significantly to 60 rpm. The satellite spin kept on decreasing with the spin rate reaching 20 RPM around 2015.

Unlike SCD-1, the SCD-2 satellite has an autonomous spin control system using magnetic actuator to keep the spin rate between 34 ± 2 rpm. When the spin rate decays to 32 rpm a maneuver is executed to increase the rotation to 36 rpm [5] [6] [7] [8] [9]. Until now, the SCD-2 attitude control system is working according to the requirement (see figure 6).

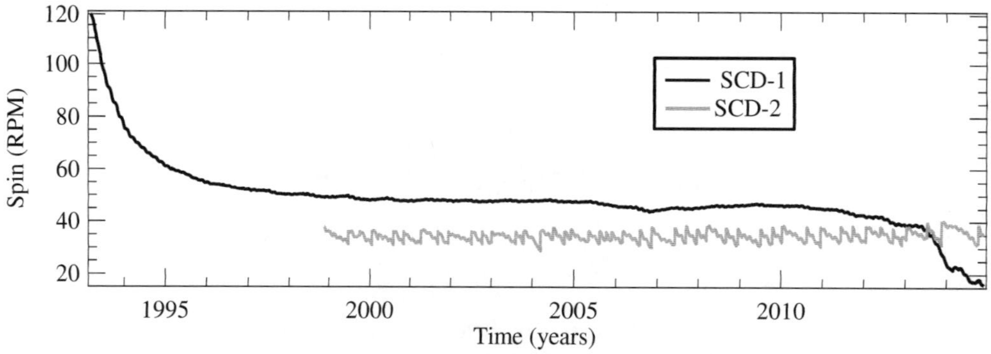

Figure 6. Spin rates of SCD-1 and SCD-2, SCC data.

In the inertial reference system, the satellites longitudinal (spin) axis are oriented based on declination and right ascension angles (figures 7 and 8). Owing to on-orbit acting perturbations, the spin axis is drifting continuously. Due to the SCD-1 ground control system, the spin axis is

IOP Conf. Series: Journal of Physics: Conf. Series **911** (2017) 012015 doi:10.1088/1742-6596/911/1/012015

maneuvered whenever the sun rays reach the lower surface of the satellite structure (thermal requirement for satellite heat dissipation). The solar aspect angle is therefore maintained in a range between 60° to 90° (figure 9). The SCD-1 maneuvers can be visualized in figure 7 by several jumps on declination angle.

The spin axis of SCD-2 has a different requirement of being perpendicular to the ecliptic plane (Sun plane), and therefore their attitude angles (figure 7 and 8) are kept in a range (declination between 40° and 80°, right ascension between 200° and 300°) corresponding to a solar aspect angle between 80° to 100°, with mean value of 90° (figure 9). The spin axis maneuvers are planned and executed by SCC, which monitors and tracks both satellites [5] [7].

Related to the RAAN and spin evolution, there is the eddy current parameter [7]. Eddy currents flowing inside the satellite coupled with the local geomagnetic field causes perturbations. Such perturbing torque is modelled by the eddy current parameter. Thus, its variation translates to major or minor perturbation on the spin axis mainly on the decay of the spin rate. In SCD-1 the eddy current parameter presented significant changes (from 800 to 0), in the period of 1993 to 1995, and from 0 to 1000 in the years 2013 to 2015 (figure 10). The parameter changes are related to spin rate decays in the same intervals of time (figures 6). In the case of the SCD-2, the eddy current parameter has values from 1300 to -500 reflecting more intense electronics activity onboard (figure 10).

The onboard core magnetic coils generate the magnetic torques to control the spin axis and to assure the compliance of requirements as far as the solar aspect angle is concerned. In figure 11, the switch-on of the magnetic coils are depicted showing the maneuvers executed along time. More than 46 maneuvers were performed for SCD-1 and more than 56 for SCD-2 thanks to the SCC ground attitude control system.

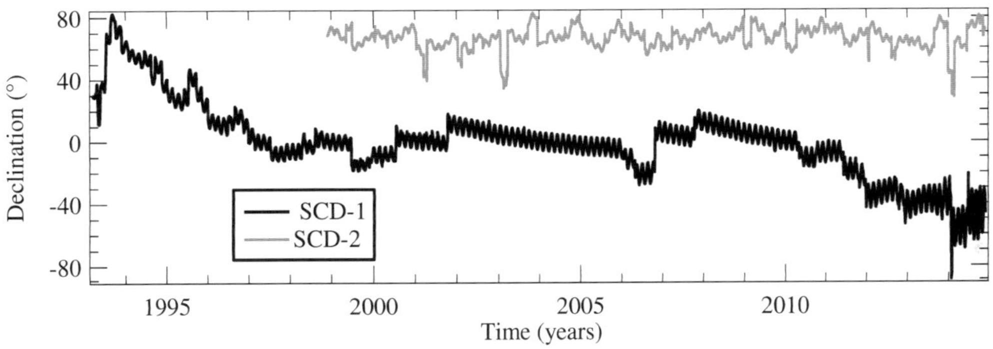

Figure 7. Declination, SCC data.

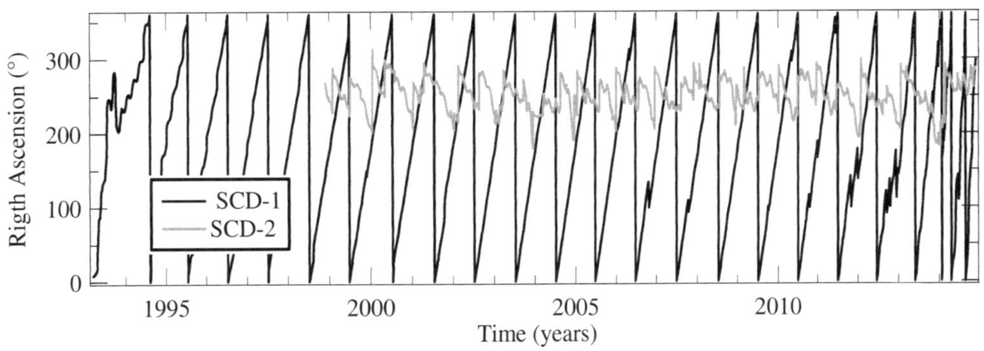

Figure 8. Right Ascension, SCC data.

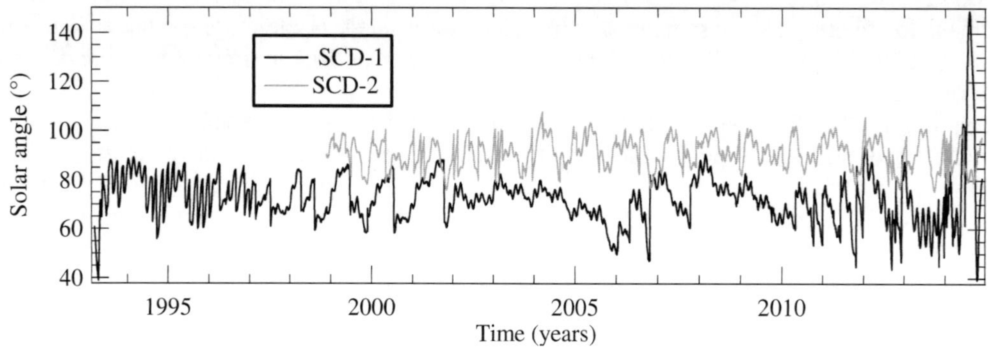

Figure 9. Solar angle, SCC data.

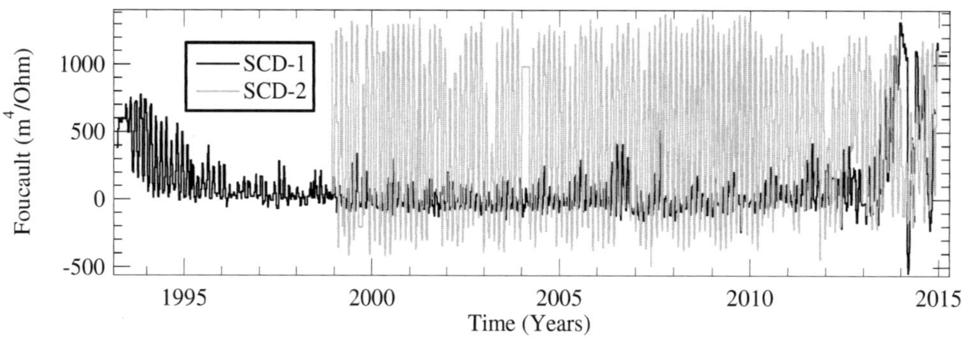

Figure 10. Eddy current parameter, SCC data.

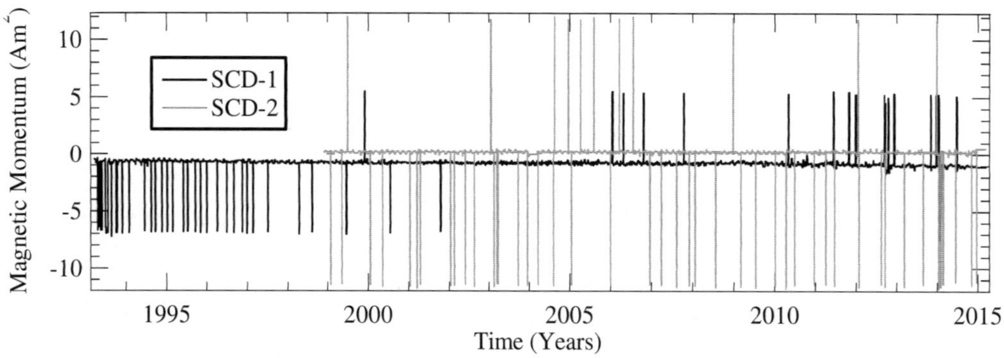

Figure 11. Magnetic momentum, SCC data.

4. Orbital decay

One important feature in satellites lifetime is the decay rate and the reentry predictions. The NASA´s Debris Assessment Software (DAS) [10] was used to estimate the satellites semi-major axis decay in the next years. According to the simulations, in a time span superior to 110 years, the apogee altitude will decay only around 200 km (altitude higher than 500 km), and both the satellites will be still in orbit, therefore becoming space debris (if they will be decommissioned before the upcoming 110 years). Such simulations show that, due to the lower area/mass ratio (around 0.01 m^2/kg) and the high initial apogee, the reentry epoch (altitude less than 120km) is not foreseeable for the next century. In figure 1, from the flight data, it is possible to see the relation between the altitude decay and the maximum solar activity. This same behaviour is output by the NASA-DAS software.

5. Conclusions

For more than 20 and 16 years in service, respectively, SCD-1 and SCD-2 satellites have shown reliable operation. The rate of change of their orbital elements is small and the main dissipative orbit perturbations acting on the satellites are the sun radiation pressure and atmospheric drag in a less extent. Estimates obtained by simulations show that the satellites will reach the highest decay around the year 2030. The satellites reentry is estimated to occur well after the year 2120.

In the attitude case, with respect to the solar aspect angle, both satellites still feature spin axis attitude angles within the operational values. In the case of SDC-2, the satellite presents a smoother attitude (less drift) than SCD-1, because of the presence of an onboard spin rate control system. As such, the spin rate of the SCD-2 remains always within operational limits whereas SCD-1 has its spin rate steadily decreasing which weakens the gyroscopic rigidness effect and yields higher drift.

In short, spin rate, declination angle, right ascension angle and sun aspect angle have shown fine behaviour during the operational life of SCD-2, because of the presence of an attitude control system which is superior to the SCD-1 system. This constitutes an evidence of the technological improvement of SCD-2 satellite compared to SCD-1.

The maximum operational time was designed, at the launching time, to be around 2 years for each satellite, but the excellent Brazilian manufacturing and engineering made it possible for the satellites to be still operating and collecting data, and to the date there are no signs of any fatal malfunction which could decommission the satellites.

Acknowledgements

The authors thank the "Instituto Nacional de Pesquisas Espaciais" (INPE), "Coordenação de Aperfeiçoamento de Pessoal de Nível Superior" (CAPES), the Science, Technology & Innovation Ministry (MCTI), and Dr. Valcir Orlando from the INPE´s Satellites Control Center for kindly providing SCD´s data.

6. References

[1] Fonseca I. M. and Bainum P. 2006 The Brazilian Satellite Program - A Survey *Adv. Astronaut. Sci.* 122 pp 505-531

[2] Vilhena R. V., Kuga H. K. and Campos D. 2002 Orbital Propagation for Brazilian Satellites Using NORAD Models *Adv. Space. Res.* **30** 2 pp 331-335

[3] Jacchia L. 1971 Semiannual variation in the heterosphere: A reappraisal *J. Geophys. Res.* **76** 19 pp 4602-4607

[4] Jacchia L. 1977 Thermospheric Temperature, Density and Composition: New Models *SAO Spec. Rep.* 375

[5] Orlando V., Kuga H. K. and Guedes U. T. V. 1998 Flight Dynamics LEOP and Routine Operations for SCD2, the INPES´s Second Environmental Data Collecting Satellite. *Spaceflight dynamics, Advances in Astronautical Sciences* (AAS-98-382), p. 1003

[6] Orlando V., Lopes R. V. F. and Kuga H. K. 1997 INPE´s Flight Dynamics Team Experience Through Four Years of SCD1 in Orbit Operation: Main Issues, Improvements and Trends. Proceedings of ISSFD (Report ESA97P)

[7] Rozenfeld P., Orlando V. and Bastos R. 1996 SCD1 Three years in orbit operation *ESA Rep. 1996 ESASP-394-1442R*

[8] Fernandez J. and Santana C. 1993 Overview of the SCD1/Pegasus Mission *Proceedings of the AIAA/USU Conference on Small Satellites, Mission Lessons, SSC7-VII-1.* Utah, EEUU.

[9] Kuga H. K., Orlando V. and Lopes, R. V. F. 1999 Flight Dynamics Operations During LEOP for the INPE´s Second Environmental Data Collecting Satellite SCD2, *RBCM – Brazilian Journal of Mechanical Sciences.* V. 21, pp 339-344

[10] NASA 2012 "Debris Assessment Software User´s Guide," Houston, TX, Lyndon B. Johnson Space Center

IOP Conf. Series: Journal of Physics: Conf. Series **911** (2017) 012016 doi:10.1088/1742-6596/911/1/012016

Reviewed plan of the ALR, the laser rangefinder for the ASTER deep space mission to the triple asteroid 2001-SN263

A G V de Brum[1], F C da Cruz[2]

[1] Universidade Federal do ABC – UFABC, Santo André, SP, Brazil
[2] Universidade de Campinas – UNICAMP/IFGW, Campinas, SP, Brazil

E-mail: antonio.brum@ufabc.edu.br

Abstract: The Brazilian deep space mission ASTER plans to send a small spacecraft to investigate the triple asteroid 2001-SN263. The launch is expected to occur in June, 2020. The main motivation of the mission is the development of technology and expertise to leverage the national space sector. The main scientific goal is the investigation of the triple asteroid 2001-SN263. To aggregate the widest possible Brazilian involvement in all mission sectors, some instruments for the mission are planned to be developed in partnership among Brazilian institutions. In this effort, a preliminary design of a laser altimeter to meet the mission needs was created and presented in 2010-2011. Since then, many studies were conducted and new information on the target asteroid was gathered. To take this into account, a review of that plan was conducted and the current results for the instrument design parameters are presented.

1. Introduction

The Brazilian deep space mission ASTER plans to send a small spacecraft to encounter and investigate the triple asteroid system 2001-SN263. There is a launch opportunity in June, 2020, with insertion in heliocentric orbit in Feb, 2021, arrival in April, 2022, time of flight: 1.85 years, exploration phase: 4 months, and estimated budget of 60 Million USD. The use of the low-cost Finnish-Russian platform MetNet is planned.

The main motivation of this mission is the development of technology and expertise to leverage the national space sector. In terms of scientific goals, the investigation should bring some indications on the formation of this triple system and each of its main three asteroids will be separately investigated as to dynamic and orbital properties, shape, size, volume, mass distribution, mineral composition, topography and surface texture, gravitational field and rotation speed.

In order to fulfill these goals, different scientific instruments are planned including one laser altimeter. The scientific data will be sent to Earth via telemetry. References [1, 2, 8, 9, 11] provide more details on the ASTER mission and the target system.

A preliminary design of a laser altimeter to meet the mission needs, named ALR (ASTER Laser Rangefinder), was created and presented [6]. Since then, many studies were conducted, including the creation of simulation software for the better understanding of the instrument operation (***ALR_Sim***®) [4, 5], new decisions were made and new information on the target asteroid was gathered. To take this into consideration, a review of that plan was conducted and the current results for the instrument design parameters are presented.

2. ALR – The Laser altimeter for space mission ASTER

Concept - The instrument measures the time of flight (T.O.F.) of the laser pulse, i. e., the time the laser pulse takes to leave the instrument, reflect on the target asteroid surface and return to the instrument. Thus, the distance and relative velocity between the spacecraft and the asteroid system can be precisely measured. A topographic profile of the target, corresponding to the course of the beam over the surface, can be produced and, by interpolation, a global model of the asteroid shape can be obtained.

Functions - The ALR will contribute with the geodesic and geophysical characterization of the asteroids that are part of the triple system. The data acquired with the equipment will be combined with data from the imaging cameras to gather up topographic features of the targets.

The instrument will also be useful in the probe navigation in the approximation stage – distance of 30 to 40 km from the center of gravity (CG) of the system – and proximity stages (to each of the asteroids; decision of reaching distances <30 km), providing precise information concerning the distance spacecraft-target, and the relative velocity between them. Also, the ALR will be used to calibrate the infrared spectrometer on board.

ALR Scientific Objectives

- A model of the shape of the main asteroids in the system should be obtained with ≤10m precision, in terms of spatial resolution (surface) and height (topography), in relation with the CG of the asteroid, in illuminated regions or those with no illumination.
- The mass of the asteroids should be defined (level of precision to be determined).
- Gravitational terms J1 and J2, which describe how the mass of the asteroid is distributed, should be determined (level of precision to be determined).

Use of the data obtained by the ALR

- Construction of topographic profiles of the asteroids;
- Creation of a model of the global shape of the asteroids;
- Support for the study of geodesic parameters of the asteroids (coordinate system, rotation, etc.);
- Support for the definition of the orbit and the modelling of gravitational data;
- Support for spacecraft maneuvers;
- Measuring surface roughness and albedo (at laser frequency);
- Support for autonomous navigation (as a sensor) at the approximation stage.

3. Design General Directions

- Design aiming to minimize instrument size and weight.
  Minimize telemetry commands and functions; simple commands: on, off, start of operation, end of operation, gain adjustments, etc. Table 1 summarizes the wanted features for the instrument.
- Use of simple and independent optics: transmitter and receptor.
  The classic Cassegrain telescope, with optical deposition and a light material mirror, but also hard and resistant (silicon carbide or aluminum).
- Low-power Electronics.

4. About the instrument operation

From initialization, the equipment should operate continuously throughout the whole investigation period of the mission (T≤6 months). The initialization should occur at a distance of less than 50 km from the target. Initially, it may be useful for navigation providing precise information concerning distance and relative velocity. Following the preliminary investigations results, the signal to start the maneuvers and obtain a closer position to the system will be given. In this case, a new spacecraft-target distance (CG of the triple system) will be defined by the mission management. References [2, 3] show more information on the physical data of the triple asteroid system.

Between 20 to 30 km, ALR can be used to investigate the dynamic characteristics of the system, as well as those related to shape, dimension and topography of the asteroids, with distance measurements error <10 m. The first scientific use of ALR should be about this distance from the main target, asteroid Alpha. In this case, the beam divergence, α (half-cone angle), should be small, and the power should be enough to obtain a return signal that is related to a 10 to 20 m footprint (radius of the illuminated area). This is sufficient to conduct the scientific analyses related to this approximation stage. For the stage of closest approximation to asteroid Alpha, about 10 km, the footprint radius should be between 5 and 10 m, which implies a beam divergence angle between 0.028° and 0.057°, in which $\alpha = \text{atan}\,(d/D)$, with d=footprint radius and D=distance to the target; 0.028°=500 μrad. The footprint is planned in order to enable the best possible mapping of the target surface. The distance between successive measurements, as a function of the shootings frequency and of the spacecraft-target relative motion (including the rotation of the target), cannot exceed a maximum value that would make it impracticable to obtain an improved surface model in terms of horizontal resolution.

Concept of the joint operation

The laser electronics controls the laser operation and directs pumping diodes inside the laser head, in which the laser beam is generated. The laser beam is collimated and expanded by the beam expander; then, it leaves the instrument through the exit optics of the transmitter, and goes towards the surface of the target asteroid. The signal reflected by the target is received by the reception telescope (receptor) and focused on the detector plan. The rangefinder has algorithms implemented in the receptor electronics to sample, record and transmit the return signal (internally, to the on board computer; externally, to Earth via telemetry).

5. Transmitter and receptor

The transmitter will be operated by shooting commands programmed in the control unit of the equipment. Operation can also be externally commanded, via telemetry. Triggering the transmitter or not will depend on the signal sent by the Attitude and Orbit Control Subsystem (AOCS), which can confirm the correct spacecraft pointing. In operation, ALR will send the collected information to a data processing and storage unit. Such data should be used on board (by AOCS, as information concerning navigation and support to the operation of other scientific instruments) and/or sent to telemetry.

The laser transmitter consists of a Nd: YAG laser, diode-pumped, Q-switched, operating at 1.064 μm wavelength. The technology of this laser is mature and can produce high energy pulses that meet the criteria of low power consumption and reduced size. The operation in the close infrared gives this laser a wide variety of efficient optics and detectors that are commercially available, which positively influences the decision to use it. The optical diode-pumped laser assures the long-life low power consumption of typically 50,000 hours or up to 18 million 10 pps shooting. A unit of laser diode operating at 808 nm will basically consist of its current source and thermoelectric coolers to stabilize the emission wavelength. Pumping geometry will be linear, including the diode laser attached to the Nd: YAG crystal, two infrared mirrors and a $LiNbO_3$ electro-optic modulator for the Q-switch.

The laser source operating in Q-switching mode should be able to produce minimum energy pulses of 5-16 mJ, lasting up to 15 ns with repetition rates varying from 1 to 10 pps, thus producing laser pulses with 1 MW of optic power. The energy of the laser pulse should be high, because, according to [10], 2001-SN263 is a C-type asteroid, which means its albedo should be relatively low at 1.064 μm. Besides, the repetition rate influences the horizontal resolution along the beam path over the asteroid surface. However, the commitment between the horizontal resolution, energy consumption and dissipation of heat was used to establish a value for the repetition rate of the order of 1 pps. Beam divergence and its pattern are mainly determined by the geometry of the optical cavity and have an impact on the size of the footprint over the asteroid terrain.

Receptor operation will be controlled by a temporal gate, which is also internally programmed in the control unit. Generally, the gate only triggers the receptor within the time gap in which the return pulse is expected. Thus, the time to gather random noise decreases, and the relation signal/noise is maximized. This time gap defines de maximum detectable altitude gap in a single shot. In its simplest operation mode, the transmitter is commanded to generate an exit pulse, while a start signal is sent to the counting clock. When the echo or return pulse is received by the receptor, a stop signal is sent to the counting clock to allow range calculation. However, when the record of the return pulse waveform is also demanded, then the return pulse is received, first pre-processed (to extract range), then sampled (at 5ns sampling frequency). The sampled return wave is then recorded and stored for further submission. The entire sampling is commanded by the receptor oscillator. Preliminary results, related to the distance to the surface, are made available for on board use (navigation). The recorded and stored waveforms are sent to Earth to be post processed in order to extract additional information like inclinations, topography details, and albedo. The registers in the clock are then formatted for the next reading.

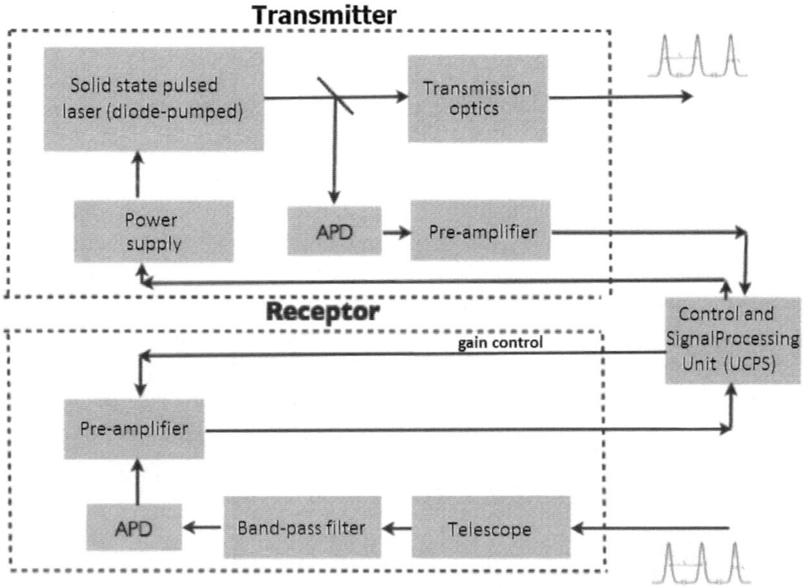

Figure 1 – Simplified block diagram of the ALR. The LIDAR, on board Hayabusa mission [12], is the main source of inspiration for the design of the ALR.

As to the instrument optics, selected the independent approach, the transmitter optics is simplified and has functions such as collimation and expansion of the exit beam according to the specification of the mission (divergence <500 μrad). A Galilean telescope with proper filters and lens can be used for this purpose. A Cassegrain type telescope is devised for the receptor, with both internal mirrors made of a hard and resistant material, such as silicon carbide (SiC) or aluminum, the latter being much lighter. The choice for aluminum requires a nonthermal design, like the one used in the laser rangefinder that flew in the NEAR-Shoemaker mission [7].

6. Electronics

According to the guidelines of the mission, it is necessary to optimize resources. In order to meet this objective, the same electronic unit should control both the transmitter and the receptor (integrated electronics). The transmitter electronics includes feeding, triggering and shooting of the laser source, besides the electronics that is dedicated to controlling the operation. The receptor electronics is in charge of its operation, which includes the photo detector element (PD) and the preamplifier, besides the control of the receptor (integrated with the transmitter). This item also detects the pulse reflected by the target, pre-processes it, obtains the shape of the received pulse wave, processes this signal, storages and sends data to other equipment on board or to Earth, via telemetry (use of memory), among others. Internal algorithms are needed here to do this processing.

7. Software

Internal algorithms - After being implemented in the electronic unit, they are used to first pre-process the signal, obtain the shape of the wave, analyze the pulse, detect the border, calculate the distance, etc. Algorithms are implemented in the electronic control unit and conduct the processing of the received signal/pulse, providing the desired information – calculation of the distance to the target, for example, and other items to be conducted in real time for use on board (other devices/instruments), or to send to Earth.

Operation modes - The device shall remain in standby mode. An auto-checking and calibration mode, integrated or not with a desirable FDIR mode (fault detection, isolation and recovery) is required.

Mode 1: navigation - In this mode, at the encounter phase (D $\leq$ 50 km), the ALR shall be triggered. Its use will solely help navigation, offering AOCS with range measurements to control proximity and relative velocity (at 1Hz frequency).

Mode 2: surface explorer – to be activated only at the stage of greater proximity with the target asteroid (within the operating range of the unit), the ALR system will be used in continuous operation, to survey of topographic features of each area to explore. Here, in addition to the measured distance, the returning pulse waveform is sampled (5 samples, 5ns intervals, 8 levels of signal intensity). Each of these campaigns can last from weeks to months.

A group of algorithms to be used in the processing and control unit of the instrument have already been the subject of study. References [4, 5] describe these algorithms along with the software created to simulate the instrument operation. The simulator software package is called ***ALR_Sim®*** and it constitutes an integral part of the instrument design. It was created with the overall objective to serve as a laboratory for the understanding, implementation and testing of all the concepts involved in distance detection using a pulsed laser rangefinder equipment based on the time of flight of the light emitted. The modeling and simulations made it possible to understand the details of the operation of an instrument similar to the ALR. Such understanding allows the identification and definition of design parameters essential to the operation, as well as the refinement of previously established values for some parameters. The Simulator provides the appropriate environment for the implementation and testing of the routines that integrate the control and signal processing software of the instrument. Additionally, the results of the simulations offered important information for the definition of specific components suitable for the composition of the unit, bearing in mind the objectives of the unit within the mission.

Table 1. Summary of predetermined features for the ASTER Laser Rangefinder.

Mass	Reduced (< 5 kg)
Dimensions	Reduced (Max: 37.5 x 23.0 x 35.0) cm^3
Power	$\leq$ 20 W
Operational Range	30 m $\leq$ D $\leq$50 km
Precision	Vertical: 10 m (D < 50 km) to 1 m (D $\leq$ 10 km) Horizontal: < 10 m
Wavelength	1.064 μm
Pulse-width	10 ns
Repetition rate (frequency)	1 Hz = > fixed. Should suffice for topographic analysis
Pulse energy	5 a 16 mJ
Transmitter's divergence	<500 μrad
Life period	< 6 months (after initialization)
Footprint	< 10 m (radius; 10 km away)
Temperature	±2°C (depending on the detector)
Self-calibration during flight	Yes
Sampling of return waveforms	Yes
Acquisition data rate	$\geq$ 12 bytes/sec

Data acquisition rate when in operation (mode 2) - As the vertical resolution of the detector has been set at 5ns, this is the returning pulse waveform sampling frequency. Therefore, a data acquisition rate of about 12 bytes/sec (1 for shooting, 2 for arrival, 5 of the waveform, other identification data) is provided for.

8. Conclusions and comments

In general terms, the most important item for the ALR design were defined and presented. This development is centralized in the Aerospace Engineering group of UFABC.

The development will be performed in stages (A, B, C, D). Stage A is at its end and it consists of the creation of a detailed development plan. At other stages, the models are created and tested. The first model, to test the functionality of the core instrument (the setting of unqualified elements or modules on a plate that is different from the final version), is in progress.

The ASTER Project has been recently inserted in the regular activities of the National Institute for Space Research (INPE).

References

[1] Araujo R A N de, Winter O C, Prado A F B de A, Sukhanov A A 2012 Stability regions around the components of the triple system 2001 SN263. Monthly Notices of the Royal Astronomical Society, **423**(4): 3058-3073. doi:10.1111/j.1365-2966.2012.21101.x.

[2] Becker T, Howell E, Nolan M, *et al* 2015 Physical modeling of triple near-Earth Asteroid (153591) 2001 SN263 from radar and optical light curve observations. Icarus **248** (2015): 499–515. doi: 10.1016/j.icarus.2014.10.04

[3] Becker T, Howell E S, Nolan M C, Magri C 2008 Physical Modeling of Triple Near-Earth Asteroid 153591(2001 SN263). Bulletin of the Am Astronomical Soc, **40**: 437.

[4] Brum A G V de, Hetem Jr., Cruz F C da, Rodrigues A P 2015(a) ALR - A Laser Altimeter for the First Brazilian Deep Space Mission. Modeling and Simulation of the Instrument and its Operation. Journal: Comp and Applied Math. **34**(2): 557-569. doi: 10.1007/s40314-014-0145-8.

[5] Brum, A G V de, Hetem Jr, Cruz F C da 2015(b) ALR – Laser altimeter for the ASTER deep space mission. Simulated operation above a surface with crater. J. Phys.: Conf. Ser. **641** 012007. doi: 10.1088/1742-6596/641/1/012007.

[6] Brum A G V de, Hetem Jr. A, Rêgo I S, Cruz F C da, Francisco C P F *et al* 2011 Preliminary Development Plan of the ALR, the Laser Rangefinder for the ASTER Deep Space Mission to the 2001 SN263 Asteroid. Journal of Aerospace Tech and Management - JATM. **3**(3): 331-338. doi: 10.5028/jatm.2011.03033611

[7] Cole T D 1998 NEAR Laser Rangefinder: A tool for the mapping and topologic study of Asteroid 433 Eros. JHU/APL Tech. Dig., **19** (2): 142-157.

[8] Macau E E N, Winter O C, Velho H F C, Sukanov A A *et al* 2011 The ASTER mission: exploring for the first time a triple system asteroid. Proc 62nd Int Astr Cong. On line at: <http://www.iafastro.net/iac/archive/browse/IAC-11/B4/2/10263/>. Accessed in Feb, 2017.

[9] Perna D, Alvarez-Candal A, Fornasier S *et al* 2014 The triple near-Earth asteroid (153591) 2001 SN263: an ultra-blue, primitive target for the Aster space mission. Astronomy & Astrophysics, **568**, id. L6, 4 pp. doi: 10.1051/0004-6361/201424447.

[10] Reddy V, Gaffey M J, Schaal M, Takir D 2008 Physical Characterization of First Triplet Near-Earth Asteroid (153591) 2001 SN263. Asteroids, Comets, Meteors 2008 held July 14-18, 2008. In Baltimore, Maryland. LPI Contribution No. 1405, paper id. 8244

[11] Sukhanov A A, Velho H F de C, Macau E E, Winter O C 2010 The Aster project: Flight to a near-Earth asteroid. Cosmic Research, **48**(5): 443-450. doi:10.1134/S0010952510050114.

[12] Tsuno K, Okumura E, Katsuyama Y *et al* 2006 Lidar on board asteroid explorer Hayabusa. Proc. 6th Int. Conf. on Space Optics, ESTEC, Noordwijk, The Netherlands (ESA SP-621, June 2006).

Colisional Cloud Debris and Propelled Evasive Maneuvers

L. S. Ferreira[1], A.D.C. Jesus[1], T.C.F. Carvalho[1], R.R.Sousa[2]

[1] UEFS, Universidade Estadual de Feira de Santana, Feira de Santana - BA, Brasil.
[2] UNESP, Universidade Estadual Paulista Júlio Mesquita Filho. Guaratinguetá,SP. Brasil.

E-mail: lsfferreira@live.com

Abstract. Space debris clouds exist at various altitudes in the environment outside the Earth. Fragmentation of debris and/or collision between the debris of a cloud increases the amount of debris, producing smaller debris. This event also increases significantly the chances of collision with operational vehicles in orbit. In this work we study clouds of debris that are close to a spacecraft in relation to its distance from the center of the Earth. The results show several layers of colliding debris depending on their size over time of evasive maneuvers of the vehicle. In addition, we have tested such maneuvers for propulsion systems with a linear and exponential mass variation model. The results show that the linear propulsion system is more efficient.

1. Introduction

Space debris are non-operational objects that orbit Earth at various altitudes. They can be captured by the gravitational field, when they come from the interior solar system, or from the outer by slingshot effect of larger bodies (Jupiter, for example) or are by-product of the space missions. According to the European Space Agency [1], since the launch of Sputnik, there have been more than 4,900 launches, which have placed about 6,600 satellites in orbit, including 3,600 remain in space, with about 1,000 operational. Most of this equipment goes into disuse and orbits the Earth. Over time, they suffer from degradations that cause propellant leakage, explosions, fragmentation, and may collide with each other due to mutual collisions and the performance of space-dissipative forces [2]. With shocks, the number of objects increases exponentially in the Earth's orbit, allowing for more frequent collisions with operational vehicles. Some of the debris is removed by atmospheric drag in LEO, but many of its layers would take decades or centuries to enter and be destroyed by friction [3]. Space cleansing has been discussed through mitigation techniques, but the alternative of reducing spatial activity or reducing the size of operating vehicles prevails. Meanwhile, in LEO, it is estimated that the average risk of catastrophic collisions is around 3% per year [4]. In a collision, the amount of energy produced is very high (~ 104 J), which would cause irreparable damage to an operational vehicle, making a mission impossible and generating more debris [5], [6], [7]. In the computational modeling of debris clouds, the number of individuals, fragmentation and their propagation must be considered to predict collisions with operational objects [2]. In this work, we propose a study of collisions between debris in a cloud that interacts with an orbiting space vehicle, without considering dissipative forces.

2. Mathematical model

The equations of the relative dynamics between a spatial debris and an operational vehicle (space colliding objects), considering them close in relation to the distance of the vehicle to the center of the

Earth can be found in 2012, [8]. The acceleration of the propulsion that will control the evasive maneuver of the object is given by the expression:

$$\ddot{a}(\mathrm{t}) = -v_e \frac{\mathrm{d}}{\mathrm{dt}} \ell n(M(t)) \tag{1}$$

$M\ (t)$ is the total mass of the vehicle (mass of the space vehicle body added to the mass of the propellant). The propellant will vary during the operation of the evasive maneuver. In this work, we will use two models for mass variation:

1) Exponential: $M(t) = m_o(\chi + e^{-\gamma t})$ and;
2) Linear: $M(t) = (M_0 + \dot{m}t)$,

with, *exponential technological parameters*, (PTE): $\gamma > 0$ - motor power factor; $\chi > 1$ - mass factor, ie, the ratio between M_o (initial vehicle mass), m_o (initial mass of the propellant) and V_e- exhaust gas velocity and, *linear technological parameters*, (PTL): $\dot{m} = dm/dt < 0$ - Rate in time change of mass, V_e - exhaust gas velocity.

Collision objects are subject to the Earth's gravitational field in LEO and the propulsion force acts only on the vehicle. In addition, we do not consider dissipative forces. The solution of the differential equations of colliding relative dynamics for a propulsion system with exponential mass variation was found in [8].

Figure 1 shows the reference system in Cartesian coordinates centered on the space vehicle. At a distance r is located a space debris or a cloud of debris. The vehicle orbits around the Earth with circular velocity, $\vec{w} = w\hat{k}$, and the y-axis is in the radius r direction.

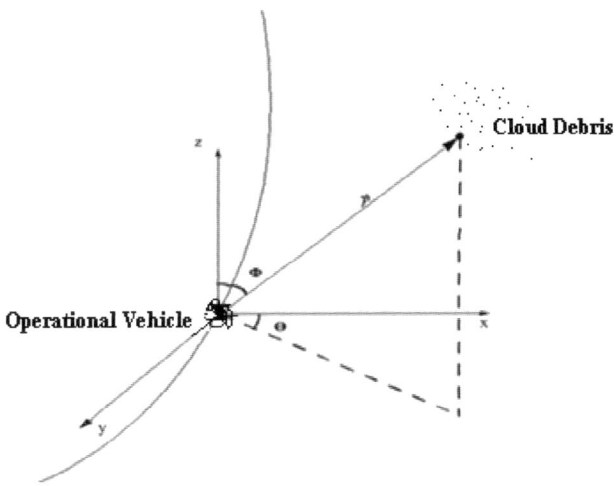

Figure 1 – Orbital Configuration and Reference System Vehicle and Cloud Space Debris

The collision between these objects occurs when their respective relative positions are null, for any values of their relative velocities, at a given time, tc. Thus, for a given pair of angles (θ, φ) there will be initial relative positions and velocities that are favorable to collision between colliding objects. In general, the operational vehicle would have the time tc to escape the collision with a spatial debris, performing an evasive maneuver.

2.1 Colisional Cloud Debris

The cloud of debris is subject to the terrestrial gravitational field and is spaced by r from the operational vehicle. The relative initial positions between the debris i and j can be written in spherical coordinates centered on the vehicle with radius equal to r, such that:

$$x_{0(i,j)} = r_0[\cos(\alpha_j)\, sin(\beta_j) - \cos(\alpha_i)\, sin(\beta_i)] \tag{2}$$

$$y_{0(i,j)} = r_0[\sin(\alpha_j)\, sin(\beta_j) - \sin(\alpha_i)\, sin(\beta_i)] \tag{3}$$

$$z_{0(i,j)} = r_0[\cos(\beta_j) - \cos(\beta_i)] \tag{4}$$

The angles α_l, β_l $(l = i, j)$ are the angular spherical coordinates of each pair (i, j) of cloud debris. The initial velocities between cloud debris are calculated using the same criteria:

$$\dot{x}_{0(i,j)} = \dot{x}_{0(j)} - \dot{x}_{0(i)} \tag{5}$$

$$\dot{y}_{0(i,j)} = \dot{y}_{0(j)} - \dot{y}_{0(i)} \tag{6}$$

$$\dot{z}_{0(i,j)} = \dot{z}_{0(j)} - \dot{z}_{0(i)} \tag{7}$$

In this way, the debris may collide with each other during its relative dynamics with the vehicle, provided that these initial conditions are favorable to the collision between them. In this work, we always combine pairs (I, j) of debris, admitting collision between only two debris at a time. The collision possibilities are obtained by combining events with each other, at each instant. If p is the number of debris that compose the cloud and n is the number of possible combinations two to two, the number of collision possibilities is:

$$n_{p,2} = \frac{p!}{2!(p-2)!} \tag{8}$$

With these equations the dynamics of the near colliding objects (debris or cloud debris and operational vehicle) can be analyzed: 1) the distribution of colliding debris in the cloud by size of debris under the action of the Earth's gravitational force; 2) the distribution of colliding debris with respect to operational vehicle, both under the action of the Earth's gravitational force and; 3) the dynamics of the operational vehicle under the action of the forces of gravity and propulsion as a measure to prevent collision with the cloud.

3.Numerical Results

In this section we present the results of the numerical simulations of the dynamics of the cloud of debris under the action of Earth's gravity. In addition we present the relative dynamics between the debris and the vehicle, subject to linear and exponential propulsions.

3.1 Colliding Cloud Debris

We simulated two clouds of 7 and 15 debris and analyzed the distribution of collisions between them subject to field gravitational Earth attraction. The clouds are close to the spacecraft at a distance far below that of the vehicle to the center of the Earth. The graphs of Figures 2 to 4 show the results of the simulations for time intervals of 2,500 s and 3,000 s. This time interval is a safe margin for the time required for the calculations and implementation of the evasive maneuver (minimum of 25 minutes).We observed that debris from a larger cloud is more likely to collide with each other. In both cases, the highest incidence of collision occurs for small and medium size debris.

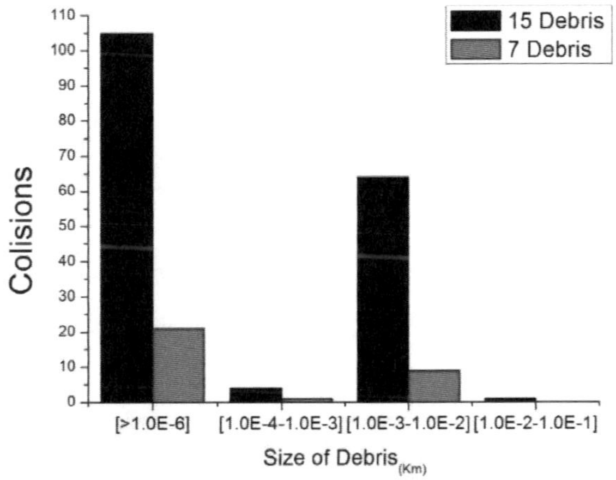

Figure 2 - Cloud Debris Collision vs. Size Colliding Debris

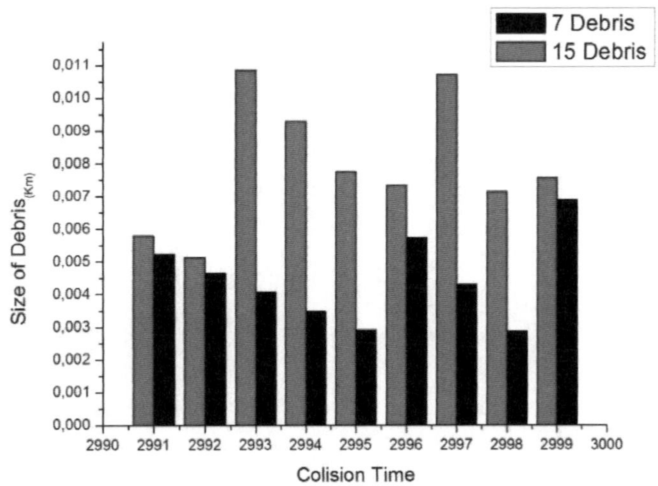

Figure 3 - Size Colliding Debris vs. Collision Time (3,000s)

This result is expected, since the number of individuals is higher in the first cloud (Figure 2). In Figure 3, we observe that the larger debris collide more (11.0 m for a cloud of 15 and 7.5 m in the cloud of 7). The collisions intensify with the increase of the collision time, under the action of the terrestrial gravitational field.

3.2 Dynamics between Debris Cloud and Vehicle

The graph of Figure 4 shows the dynamics of the 15 debris of the cloud relative to the vehicle without propulsion. The initial distance between the cloud debris and the spacecraft is 3.0 km. The D_i, i = 1,15 are the cloud debris. Table 1 shows the angular coordinates of the debris in the cloud.

IOP Conf. Series: Journal of Physics: Conf. Series **911** (2017) 012017 doi:10.1088/1742-6596/911/1/012017

TABLE 1 - Angular Coordinates - Cloud Debris

Debris	***θ***	***ϕ***
1	165.0°	23.0°
2	137.0°	38.0°
3	147.0°	157.0°
4	120.0°	120.0°
5	149.0°	140.0°
6	160.0°	80°
7	124.0°	16.0°
8	135.0°	126.0°
9	180.0°	75.0°
10	113.0°	124.0°
11	48.0°	137.0°
12	129.0°	136.0°
13	15.0°	150.0°
14	135.0°	175°
15	153.0°	135.0°

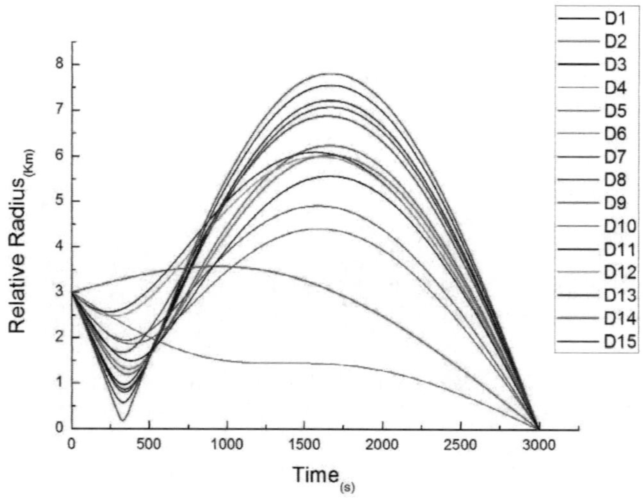

Figure 4 - Relative Final Position vs. Collision Time

In the course of time, before 500s, we observe that the debris close to the vehicle and does not collide with it. The collision occurs at 3,000s, since there is no propulsion. At this moment, a greater incidence of collision between them may occur, depending on their size. At 1,600 s greater relative distance occurs between the debris and the vehicle. Figure 5 shows this dynamics propelled.

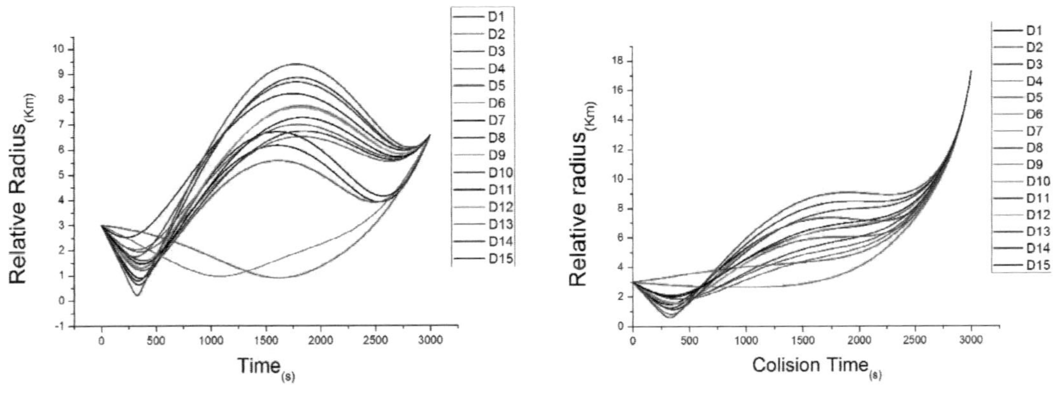

Figure 5 - Relative Final Position vs. Collision Time - Propelled: a) Exponential, b) Linear

These results show the efficiency of the linear propulsion w.r.t. exponential propulsion system in the implementation of the evasive maneuver in front of the collision with the debris cloud, because it escapes collision with relatively larger debris. We highlight in Figure 6, below, an evasive maneuver of a collision with single cloud debris under the operation of both propulsion systems. (PTL: $\dot{m} = -0.0160\ kg/s$, Mo=49.5 kg, PTE: $\gamma = 10^{-6}\ 1/s$, $\chi = 10$)

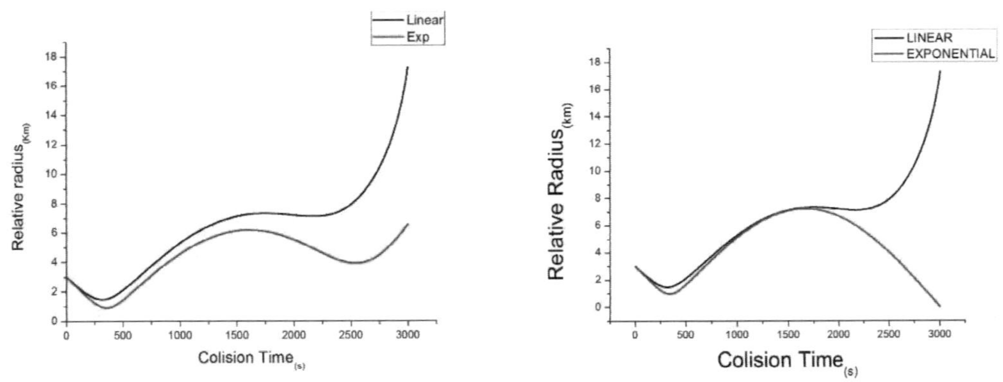

Figure 6 – Relative final position between one debris and the operational vehicle vs. Collision Time: a) PTL: V_e=0.0025 km/s, PTE: V_e=2.5 km/s. b) PTL, PTE: V_e=0.0025 km/s.

We observed that the linear propulsion system is more efficient to escape larger debris from the exponential system at any velocity regime. At low velocities (Figure 6b), the exponential system performs collision evasive maneuvers with debris up to 6.57 m, while the linear system with debris up to 18.0 km.

4. Conclusions

We study the dynamics of debris cloud near a operational vehicle in the terrestrial gravitational field. We observed that the larger the debris cloud has the greater chance of collision between those of small and medium sizes, as the time of exposure to the field increases. The evasive maneuvers with linear propulsion are more efficient to escape from growing debris and even in regimes of small velocities, in relation to the exponential propulsion. We do not consider the atmospheric drag in our approach, because it is a basic theoretical investigation to study the effect of earth's gravity on the cloud of

IOP Conf. Series: Journal of Physics: Conf. Series **911** (2017) 012017 doi:10.1088/1742-6596/911/1/012017

debris, in order to know its scattering. We expect that the action of the atmospheric drag would certainly redistribute the possibilities of collision for relative high speeds.

5. References

[1] ESA Space Debris Working Group. Space Debris, ESA SP-1109, November 1998.

[2] Zhang, Binbin; Wang, Zhaokui; Zhang, Yulin. 2016. An analytic method of space debris cloud evolution and its collision evaluation for constellation satellites. *Advances in Space Research,* **v. 58**, n. 6, pp 903-913,.

[3] Rossi A 2011 Space debris *Scholarpedia*, **6(1):10595**. Available in: http://www.scholarpedia.org/article/Space_debris. Last acess: 27/02/2014.

[4] Lafleur J M 2011.Extension of a simple mathematical model for orbital debris proliferation and mitigation.

[5] Klinkrad H 2006 Space Debris: Models and Risk Analysis. *Springer Praxis*, Berlin-Heidelberg.

[6] Rossi A Valsecchi G B 2006 Collision risk against space debris in Earth orbits. *Cel. Mech. Dyn. Astron.* **95**, 345–356

[7] Smirnov N N 2002 Space Debris Hazard Evaluation and Mitigation. *Taylor & Francis*, London-New York.

[8] Jesus A D C et al 2012 Evasive Maneuvers in Space Debris Environment and Technological Parameters. *Mathematical Problems in Engineering,* Hindawi Publishing Corporation,**v2012,** pp15

Artificial satellites orbiting planetary satellites: critical inclination and sun-synchronous orbits

Maria Lívia G T X da Costa[1], Rodolpho Vilhena de Moraes[1], Jean Paulo S Carvalho[2] and Antônio Fernando B A Prado[3]

[1] Institute of Science and Technology, Federal University of São Paulo, Brazil
[2] Federal University of Recôncavo da Bahia, Brazil
[3] National Institute for Space Research, Space Mechanics and Control Division, Brazil

E-mail: livia.thibes@gmail.com

Abstract. The behavior of critical inclinations and sun-synchronous orbits of artificial satellites orbiting planetary satellites are analyzed considering, simultaneously, the influence of the harmonics J_2 and C_{22}, both due to the non-uniform mass distribution of the natural satellite. In the present research, the central bodies of interest are the Moon and two of the Galilean moons: Io and Europa.

1. Introduction

Recent space missions show interest in the exploration of planetary satellites of our Solar System. It is interesting to note that the internal mass distribution for some of the lunar satellites differ from the Earth. This fact, in general, implies in a change of the hierarchical order of the harmonic coefficients, as is the case of the Moon (see table 1).

Table 1. Spherical harmonic coefficients.

Body	J_2 (10^{-6})	C_{22} (10^{-6})	$\sigma \triangleq J_2/C_{22}$	J_3 (10^{-6})	J_4 (10^{-6})	J_5 (10^{-6})	J_6 (10^{-6})
Moon[1]	203.237	22.35700	9.090531	8.47590	−9.59193	0.715409	−13.5777
Io[2]	1859.5	558.8	3.327666	—	—	—	—
Europa[2]	435.50	131.5	3.311787	—	—	—	—

Due to the characteristics of the gravity field of the Moon, the potential truncated up to the J_9 term produces different effects when compared to the frozen orbits which were obtained using the potential truncated up to J_7 [3, 4]. Furthermore, as an example of this effect, it can be noted in the time variation of the eccentricity that its amplitude increases as a consequence of considering higher order zonal harmonics as shown in Fig. 1. For a Moon's orbiter in an altitude of about 110 km, the amplitude of the time variation of the eccentricity decreases when the first order perturbation due to the gravitational attraction of the Earth is taken into account as well.

IOP Conf. Series: Journal of Physics: Conf. Series **911** (2017) 012018 doi:10.1088/1742-6596/911/1/012018

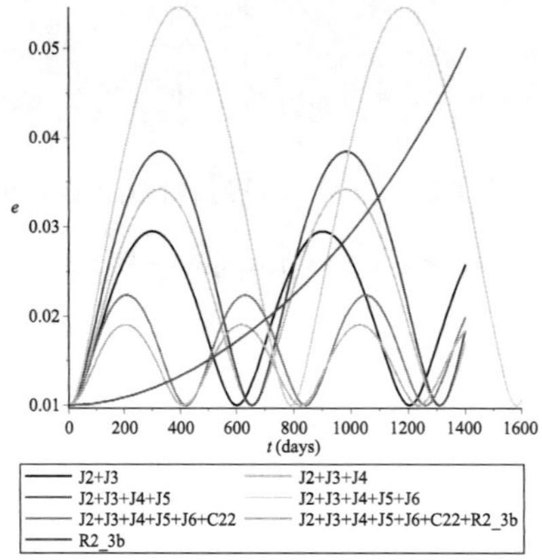

Figure 1. Time variation of the eccentricity for a Moon's satellite in a low orbit.

Critical inclination is a special value of the inclination of the orbital plane that makes the argument of the periapsis stay fixed when some perturbations are considered. This condition can be helpful when applying impulses in orbital maneuvers. Sun-synchronous orbits are those where the artificial satellite has a constant perspective on the Sun and, consequently, the orbital plane makes a constant angle with the radius Sun-satellite. Such orbits are of practical interest because the ascending node will lie at a fixed local time [5].

Since these two kinds of orbits are important and desirable for determined space missions, in the present work, a study on the orbital behavior of lunar artificial satellites at low altitude is made considering the action of perturbative forces as a consequence of the non-uniform distribution of mass of the central body, particularly, the oblateness J_2 and the equatorial ellipticity C_{22}.

2. The potential considered

According to [6], the disturbing potential thanks to the gravitational attraction of the primary body, taking into account secular terms up to J_2, long period terms up to C_{22} and eccentricity up to e^2, is expressed by

$$R = \frac{1}{8}\frac{\mu}{a^3}{a_e}^2[6J_2\cos(i)^2 - 3J_2e^2 - 2J_2 - 18C_{22}\cos(2\Omega)e^2 + 18C_{22}\cos(2\Omega)\cos(i)^2e^2 - 12C_{22}\cos(2\Omega) \quad (1)$$
$$+ 12C_{22}\cos(2\Omega)\cos(i)^2 + 9J_2\cos(i)^2e^2],$$

where μ is the gravitational parameter; a, e, i and Ω are the keplerian orbital elements and a_e is the equatorial radius of the moon orbited.

3. Critical inclination

Critical inclination $i \triangleq i_c$ is the value of the inclination that makes the time derivative of the argument of periapsis (ω) to be zero. Thus, substituting Eq. (1) in the Lagrange Planetary Equations [7] and solving for $d\omega/dt = 0$, we obtain

$$\cos(i_c)^2 = \frac{J_2 - 6C_{22}\cos(2\Omega)}{5J_2 - 10C_{22}\cos(2\Omega)}. \quad (2)$$

If C_{22} is neglected, the critical inclination for direct orbits is $i_c = 63.43°$ and for retrograde (*i.e.*, orbits with inclination between 90 and 180 degrees, so the spacecraft moves in the counterclockwise sense) ones, $i_c = 116.56°$. Otherwise, the critical inclination is a function of the ascending node Ω. If the coefficients J_2 and C_{22} are close enough to each other the critical inclination does not exist for some values of Ω. This fact is pointed by [8, 9].

For initial conditions of the ascending node in the interval $[0, 2\pi]$, Figs. 2 and 3 shows ranges of values where it is possible to find critical inclinations for spacecrafts orbiting the Moon, Io and Europa, for direct and retrograde orbits, respectively.

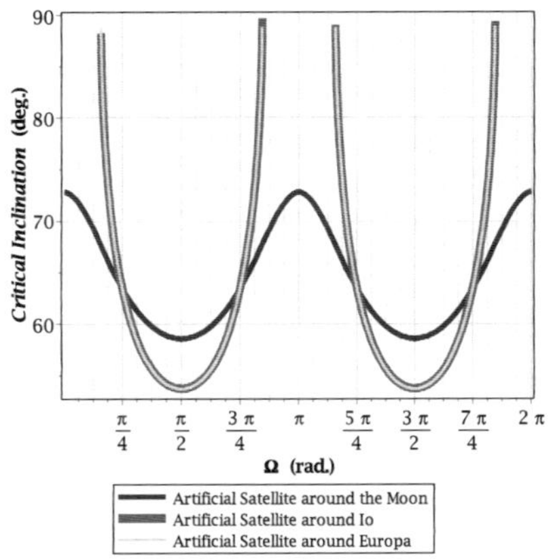

Figure 2. Critical inclinations for direct orbits.

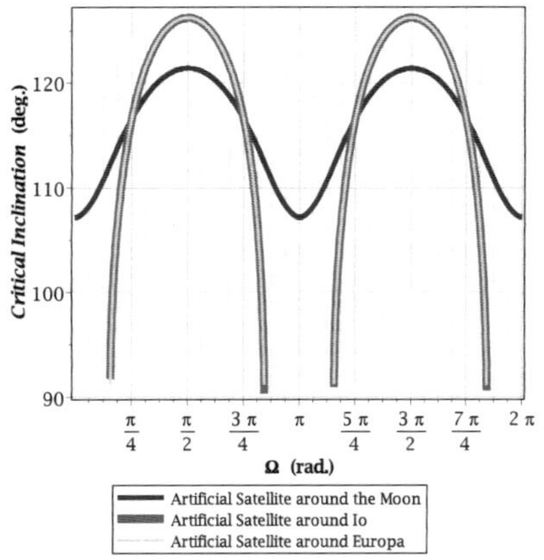

Figure 3. Critical inclinations for retrograde orbits.

Observe that the curve of the critical inclination is the same for orbits around Io and Europa. In fact, for this model, this occurs for all bodies such that the ratio $\sigma = 3.3333\bar{3}$, *i.e.*, bodies in hydrostatic equilibrium (a balance between the gravitational field and the pressure gradient). Therefore, differently of the Moon's case, the critical inclination does not exist for all values of Ω of the domain for Io and Europa.

4. Sun-synchronous orbits

For a sun-synchronous orbit to develop, the orbital plane must rotate in inertial space with the angular velocity of the central body in its orbit around the Sun, in other words, the orbital plane must precess at this rate [5]. Thus, it follows that the precession rate is given by

$$\Delta\Omega \triangleq \frac{d\Omega}{dt} = \left[\frac{T}{y}\right] 360° \; / \; lunar\ day, \tag{3}$$

where T is the orbital period of the moon around its planet and y is the orbital period of the moon around the Sun. Substituting Eq. (1) in $d\Omega/dt$ from the Lagrange Planetary Equations [5] and solving for $i \triangleq i_s$ (sun-synchronous inclination), considering terms up to the order of e^2, an expression to compute i_s is found

IOP Conf. Series: Journal of Physics: Conf. Series **911** (2017) 012018 doi:10.1088/1742-6596/911/1/012018

$$\cos(i_s) = \frac{2a^2 \Delta\Omega}{3{a_e}^2 n \left[1 + 2e^2\right] \left[J_2 + 2C_{22} \cos\left(2\Omega\right)\right]}. \tag{4}$$

For particular values of semi-major axis a and eccentricity e, the combined effect of J_2 and C_{22} on the sun-synchronous orbits around the Moon, Io and Europa are shown in Figs. 4-9.

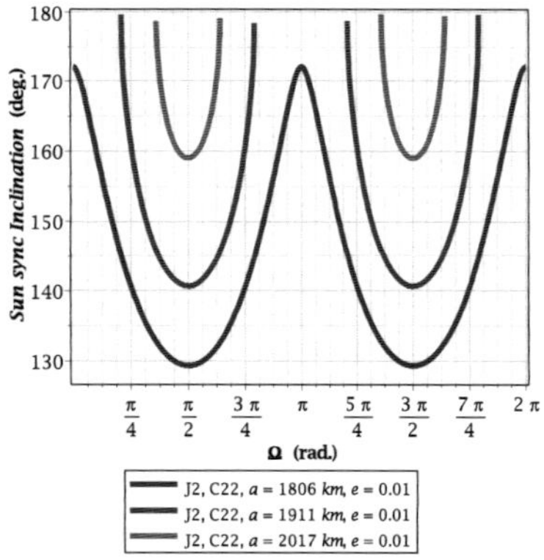

Figure 4. Sun-sync. orbits around the Moon for $e = 0.01$.

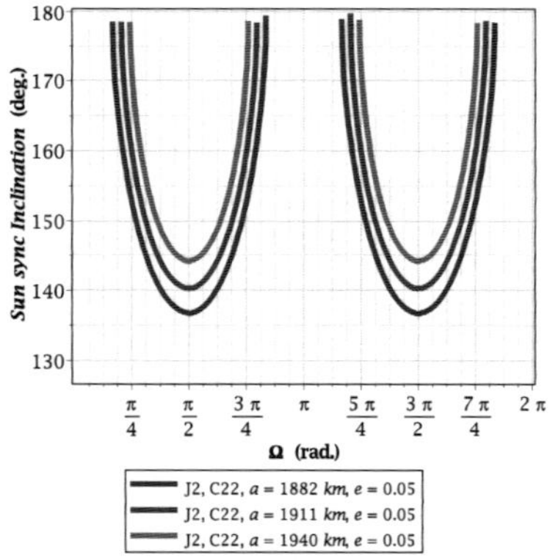

Figure 5. Sun-sync. orbits around the Moon for $e = 0.05$.

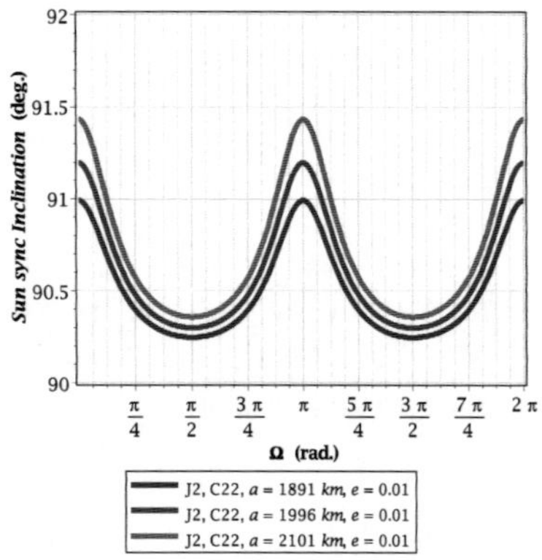

Figure 6. Sun-sync. orbits around Io for $e = 0.01$.

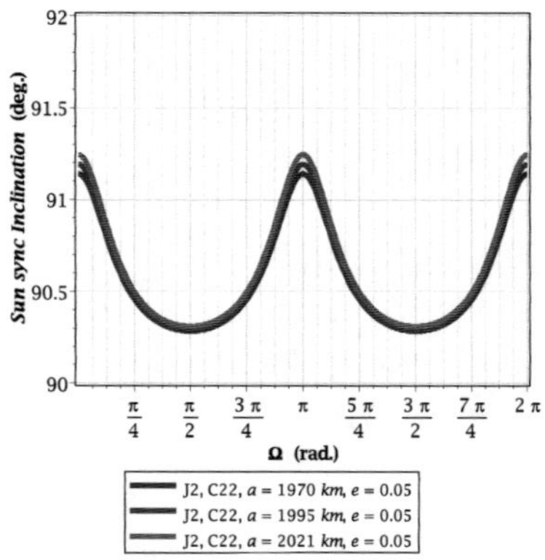

Figure 7. Sun-sync. orbits around Io for $e = 0.05$.

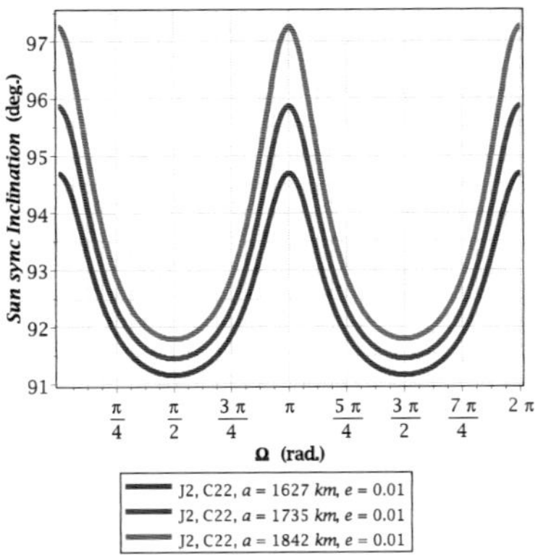

Figure 8. Sun-sync. orbits around Europa for $e = 0.01$.

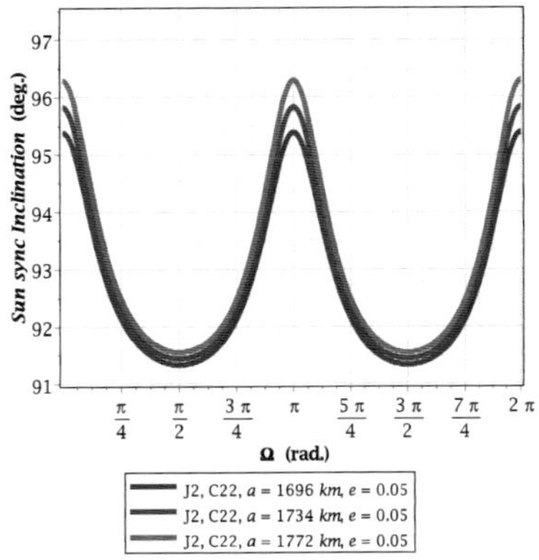

Figure 9. Sun-sync. orbits around Europa for $e = 0.05$.

It can be seen for Io and Europa that the sun-synchronous inclination exists for all the domain of the longitude of the ascending node (Ω) at the low altitude orbits analyzed as shown in Figs. 6-9. For the Moon, see Figs. 4 and 5, the behavior of the sun-synchronous inclination changes with the chosen values of the semi-major axis (a) and the eccentricity (e).

5. Conclusion

The motion of artificial satellites at low orbits are strongly affected by the central body oblateness and equatorial ellipticity. It is well known that these natural effects can be used as an advantage in the moons exploration of our Solar System. In this way, it is important to study and find critical and sun-synchronous inclinations for cases where these effects can be used as an advantage.

In the analysis of the critical inclination, it is verified the existence of direct and retrograde orbits around the Moon, Io and Europa through the Figs. 2 and 3. It is known that the critical inclination expression, see Eq. (2), is a function of the ascending node and it depends on the mass and the constants J_2 and C_{22}, which dictate the behavior of the curve by the ratio σ [8]. The larger the value of σ, the more well behaved is the curve. Otherwise, the range of ascending nodes values that does not exist critical inclination i_c is larger.

From table 1 it is seen that the ratio of the Moon is $\sigma = 9.090531$ and there are critical inclinations for all values of Ω in the domain as shown in Figs. 2 and 3. By contrast, the values of σ for Io and Europa are smaller, so the curves have discontinuities. When the central body has $\sigma = 3.3333\bar{3}$ or close enough, as is the case for Io and Europa (table 1), the curve of critical inclination is about the same. This fact can be verified in other moons with hydrostatic equilibrium in our Solar System [10].

For the case of the sun-synchronous orbits, the expression for i_s, besides depending on J_2, C_{22} and Ω, also depends on the semi-major axis a and eccentricity e of the orbit. Therefore, to find convenient sun-synchronous inclinations, beyond analysing the ratio σ, it is needed to choose proper initial conditions for a and e as well.

For the case of bodies with ratio $\sigma = 3.3333\bar{3}$ or close enough, the sun-synchronous inclinations exist for all the domain of the longitude of the ascending node at the low altitude orbits analyzed in this research, as can be seen in Figs. 6-9. This fact can be verified for other moons in hydrostatic equilibrium [10]. When σ is not close to $3.3333\bar{3}$, the behavior of the sun-synchronous inclinations change with the values of the semi-major axis and eccentricity. It may not exist for certain values of these parameters. In the Moon's case this is clearly seen, as shown in Figs. 4 and 5.

Acknowledgments
This work was supported by the National Council for Scientific and Technological Development (CNPq), process numbers 306953/2014-5, 406841/2016-0, 301338/2016-7 and by the São Paulo Research Foundation (FAPESP), process numbers 2011/08171-3, 2016/14665-2.

References
[1] Chen J Y, Ning J S, Zhang C Y and Luo J 2005 On the determination of the lunar gravity field from the first Chinese lunar prospector mission *Chinese J. Geophys.* **48** 303-11
[2] Schubert G, Anderson J D, Spohn T and McKinnon W B 2004 *Interior Composition, Structure and Dynamics of the Galilean Satellites Jupiter: The Planet, Satellites and Magnetosphere* vol 1 ed Bagenal F *et al* (Cambridge: Cambridge University Press) chapter 13 pp 281-306
[3] Vilhena de Moraes R, Carvalho J P S and Prado A F B A 2010 *Lunar Artificial Satellites: Frozen and Helio-Synchronous Orbits Proc. of the IAF Global Lunar Conf./11th ILEWG Conf. on Exploration and Utilisation of the Moon* (Beijing)
[4] Carvalho J P S, Vilhena de Moraes R and Prado A F B A 2011 Planetary satellite orbiters: applications for the Moon *Math. Probl. Eng.* **2011** Article ID 187478 doi:10.1155/2011/187478
[5] Curtis H D 2005 *Orbital Mechanics for Engineering Students* vol 1 (Oxford: Elsevier Butterworth-Heinemann) chapter 4 p 180 ISBN 0 7506 6169 0
[6] Carvalho J P S, Vilhena de Moraes R and Prado A F B A 2009 Nonsphericity of the Moon and near sun-synchronous polar lunar orbits *Math. Probl. Eng.* **2009** Article ID 740460 doi:10.1155/2009/740460
[7] Kaula W M 2013 *Theory of Satellite Geodesy: Applications of Satellites to Geodesy* vol 1 (Mineola, New York: Dover Publications) chapter 3 p 29 ISBN 978-0-486-41465-2
[8] De Saedeleer B and Henrard J 2005 The combined effect of J_2 and C_{22} on the critical inclination of a lunar orbiter *Adv Space Res* **37** 80-7
[9] Vilhena de Moraes R, Costa M L G T X, Carvalho J P S and Prado A F B A 2016 *Artificial Satellites Orbiting Planetary Satellites: Critical Inclination and Helio-Synchronous Orbits Proc. of the 67th International Astronautical Congress* (Guadalajara)
[10] Gonçalves L D, Costa M L G T X, Vilhena de Moraes R and Rocco E M 2017 *Investigation of Critical Inclination and Sun-Synchronous Orbits around the Moons of Saturn Proc. of the 27th AAS/AIAA Space Flight Mechanics Meeting* (San Antonio)

IOP Conf. Series: Journal of Physics: Conf. Series **911** (2017) 012019 doi:10.1088/1742-6596/911/1/012019

Simulation of the trajectories described by a space vehicle around the asteroid 243 Ida and its natural satellite Dactyl

E M Rocco and L D Gonçalves

National Institute for Space Research, INPE, São José dos Campos, Brazil

E-mail: evandro.rocco@inpe.br; lianadgon@gmail.com

Abstract. The asteroid 243 Ida located in the asteroid belt, between Mars and Jupiter, is the fourth largest asteroid of the Koronis asteroid family, with an average diameter of 31.3 km and a mass around 4.2×10^{16} kg, and a small moon, Dactyl. In order to study the dynamics of this system, orbital trajectories are simulated around Ida considering, besides the gravitational attraction of Dactyl, the non-central gravitational field of the asteroid, defined by a polyhedral model that defines the shape and the non-uniform mass distribution of the body. In this way, the magnitude and the behaviour of such forces, and also their influence on the orbital elements that define the trajectory of the space vehicle, are evaluated and analysed.

1. Introduction

The asteroid 243 Ida located in the asteroid belt, between Mars and Jupiter, belongs to the family of asteroids Koronis whose origin is supposed to be the result of the collision of two major bodies [1]. The asteroid Ida is the fourth largest of the Koronis family having an average diameter of 31.3 km and a mass around 4.2×10^{16} kg. This asteroid has another interesting feature: it has a small moon, called Dactyl, discovered by the Galileo spacecraft during its approach to the asteroid [2-4]. But the orbit of this small moon was not precisely determined since the images sent by Galileo were insufficient to accurately determine Dactyl's orbit around Ida, as illustrated in Figures 1 and 2.

Figure 1. Ida and Dactyl
Source: NASA/JPL [5]

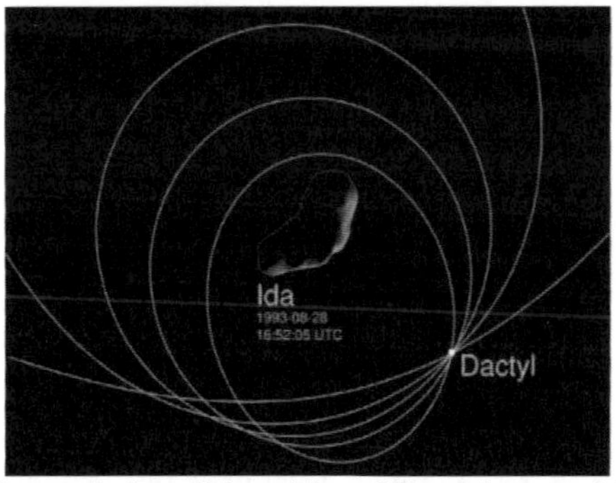

Figure 2. Possible orbits of Dactyl around Ida
Source: NASA/JPL - Dactyl potential orbits [5]

It was considered in this work the orbital motion of a spacecraft around Ida, perturbed by the gravitational attraction of Dactyl, taking into account the Dactyl orbit almost circular with radius of 90 km [6], mass of 2380 kg, and the non-central gravitational field of Ida. The gravitational field was defined by a polyhedral model of the asteroid volume, provided by NASA [7], associated to a model of mass concentrations allocated in the barycenter of the polyhedra that compose the volume of the asteroid [8-12]. Using this approach and the Spacecraft Trajectory Simulator (STRS) [13-16], it is possible to map the gravitational field around the asteroid and determine the trajectory that would be described by a spacecraft around Ida / Dactyl, which characterizes the main objective of this work.

2. Gravitational Perturbation

The gravitational field of an asteroid such as 243 Ida cannot be considered as a central force field since its shape does not resemble that of a sphere. Due to the irregular shape of the asteroid, the trajectory of a spacecraft in its orbit is perturbed by the gravitational potential generated by the asteroid. In this way the modeling of the gravitational field and the orbital perturbation imposed on the vehicle is fundamental for the simulation of the movement dynamics.

In this work we use the polyhedral model provided by NASA to model the asteroid mass distribution [7 and 17]. It was considered that in the barycenter of each polyhedron is allocated a mass concentration equivalent to the mass of the respective polyhedron. In this way, the gravitational force applied to the vehicle due to each of the mass concentrations can be calculated. By effecting the integral of all forces we obtain the resultant gravitational force. The comparison between this resultant with the gravitational force, that would be generated if the force field were central, provides the disturbing force that is applied to the vehicle at each step of the simulation [10-11 and 18], as shown in Figure 3.

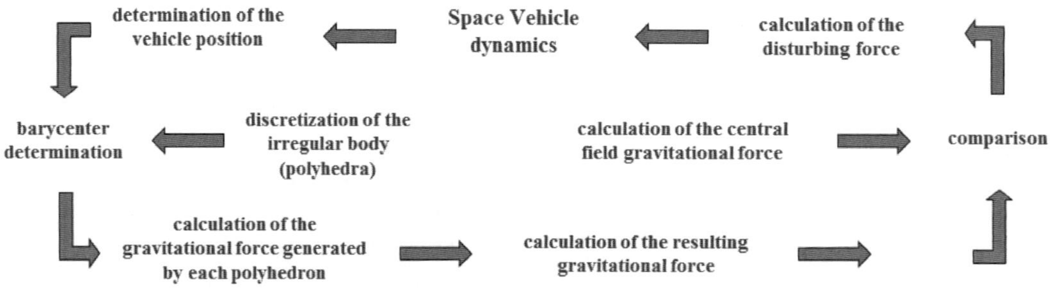

Figure 3. Simulation flowchart

The gravitational potential U of the asteroid is given by equation (1). ρ is density and $\mathrm{d}V$ is a small differential volume relative to the asteroid. Due to the complexity of the calculation of equation (1) to model the gravitational potential considering the irregular shape of the asteroid, the total volume V can be divide into n tetrahedrons, concentrating the mass M_i of each volume V_i in the barycenter of each tetrahedron, called secondary centers of attraction, as modeled in STRS and used in [8-12] and [18].

$$U = -G \int_V \frac{\rho\,\mathrm{d}V}{\|\vec{r}\|} \tag{1}$$

The gravitational force applied to the spacecraft relative to the secondary centers of attraction with mass M_i is expressed by equation (2), using a reference frame centered at the asteroid center of mass with the axes coincidences with the principal inertia axes, where G is the gravitational universal constant, m is the spacecraft mass and $\vec{r}$ is de radius vector related to the spacecraft position.

$$F_{G_{ix}}\hat{\imath} + F_{G_{iy}}\hat{\jmath} + F_{G_{iz}}\hat{k} = \left(-\frac{GmM_i}{\|\vec{r_{\imath x}}\|^2}\right)\hat{\imath} + \left(-\frac{GmM_i}{\|\vec{r_{\imath y}}\|^2}\right)\hat{\jmath} + \left(-\frac{GmM_i}{\|\vec{r_{\imath z}}\|^2}\right)\hat{k}, \quad i = 1,2,3\,\dots,n \tag{2}$$

Therefore, the resultant of the perturbing gravitational force $\overrightarrow{F_{GP}}$ applied to the spacecraft can be modeled as the difference between the force referring to the central field and the sum of the gravitational forces due to the concentrations masses, given by equation (3), where M is the total mass of the asteroid.

$$\overrightarrow{F_{GP}} = -\frac{GmM}{\|\vec{r}\|^2}\hat{r} + \sum_{i=1}^{n}\left[\left(\frac{GmM_i}{\|\overrightarrow{r_{ix}}\|^2}\right)\hat{\imath} + \left(\frac{GmM_i}{\|\overrightarrow{r_{iy}}\|^2}\right)\hat{\jmath} + \left(\frac{GmM_i}{\|\overrightarrow{r_{iz}}\|^2}\right)\hat{k}\right] \quad (3)$$

3. Simulations and results

Simulations were carried out with a duration of 30 and 100 days. The initial semi-major axis was considered to be 60 km in all simulations. Several simulations for different orbital inclinations were performed, varying the inclination from zero to 90°, but just the results referring to the inclination of 35° will be shown here. Although the orbit used is close to a circular orbit, the altitude varies significantly due to the irregular asteroid shape. The gravitational perturbation due to the irregular shape of the asteroid is a function of the position of the vehicle and the rotation of the asteroid. These parameters depend on the orbital plane adopted. The effect of the radiation pressure was taken into account considering the solar irradiance constant for a distance of 2.7 AU of 187.38 W/m^2, and the optical properties of the spacecraft surface (absorption coefficient 0.3; specular coefficient 0.3; diffuse coefficient 0.4) for a projection area of 10 m^2 in all directions. However, the graphs related to the radiation pressure were omitted since the velocity increment due to this perturbation remains approximately constant, around 10^{-5} m/s. It was considered a spacecraft with 1000 kg.

For both simulations the following initial conditions were adopted: a = 60 km; e = 0,000001; i = 35°; Ω = 0; ω = 0. The Simulation 1 considers a simulation time of 30 days and the Simulation 2 of 100 days. For the simulations 1 and 2, respectively, the figures 4 and 5 show the velocity increment applied to the vehicle considering all the perturbations (Cowell's method) for each axis (green, magenta and blue) and also for the absolute value of the disturbance (red). The variation of the perturbation observed in the figure occurs mainly due to the non-central gravitational field of Ida. Figures 6 and 7 show the velocity increment due to the Dactyl gravitational perturbation for each axis (green, magenta and blue) and also for the absolute value (red). It can be seen in the figures that occur peaks in the moments of greatest approximation with the Dactyl along the trajectory of the spacecraft.

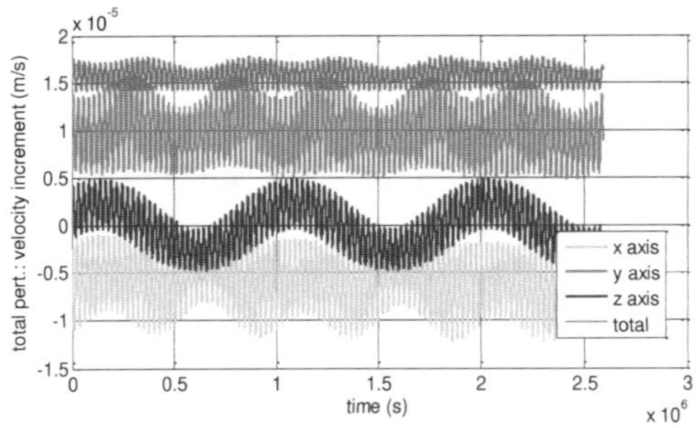

Figure 4. Total perturbation (sim. 1)

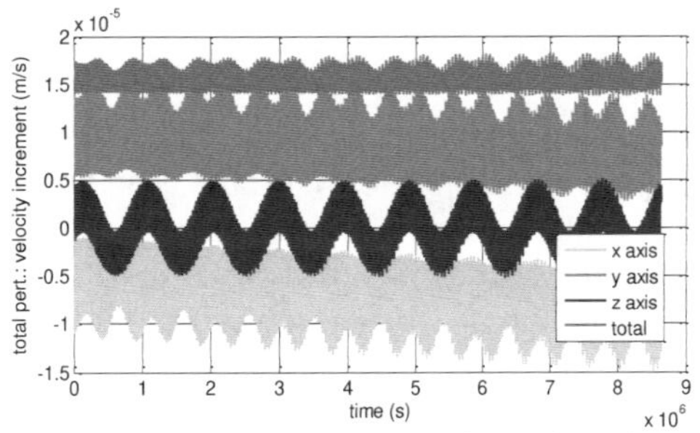

Figure 5. Total perturbation (sim. 2)

Figures 8 and 9 show the disturbing velocity increment due to the gravitational pull of the Sun, once again, for each of the axes and for the absolute value. Figures 10 and 11 show the sum of the disturbance velocity increment, due to all the combined perturbations, throughout the simulations. The sum of the disturbance velocity increment can be considered analogous to the integral of the disturbances acting on the vehicle along its trajectory, during the simulated time interval. This sum can be used as a parameter to be minimized during the analysis of a mission of a spacecraft that aims to orbit the asteroid, in order to find less disturbed trajectories that may require a lower fuel consumption to maintain the trajectory close to nominal trajectory and, in this way, enable the execution of the mission for which the vehicle was designed. Figures 12 to 19 show the variation of the orbital parameters of the spacecraft due to the action of the gravitational perturbation generated by the non-homogeneity of the mass distribution of Ida and the gravitational attractions of the Sun and Dactyl. It can be seen in figures 12 and 13 that the semi-major axis of the orbit presents short-period variations, oscillating with maximum amplitude around 25 m. However, there is a long-term trend, which is most evident in Figure 13 for the second simulation, which causes the semi-major axis to deviate more and more over time. This variation of the semi-major axis is corroborated by the variation of the eccentricity of the orbit that can be observed in figures 14 and 15. It is clear that there is a possibility of the vehicle escaping from the region of influence of Ida if the deviations in the semi-major axis and eccentricity continue to increase over time. This fact illustrates the need to find minimally disturbed trajectories by evaluating the sum of the disturbance shown in Figures 10 and 11.

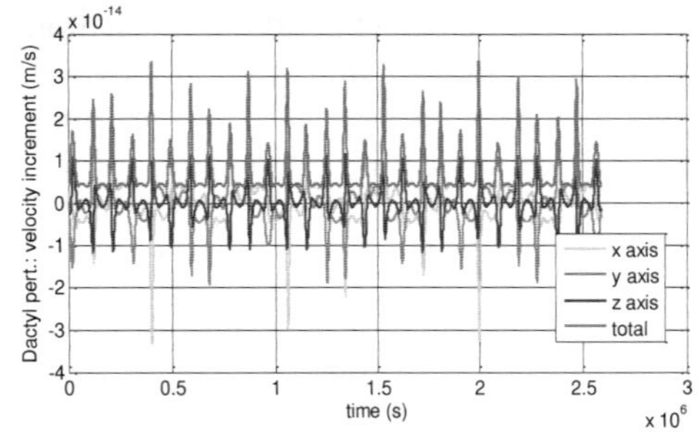

Figure 6. Dactyl perturbation (sim. 1)

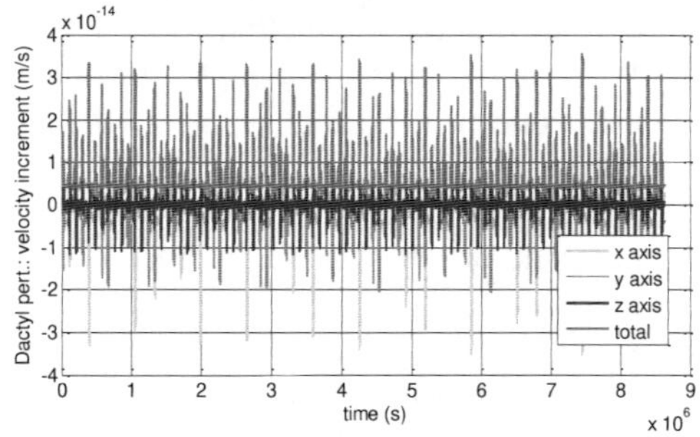

Figure 7. Dactyl perturbation (sim. 2)

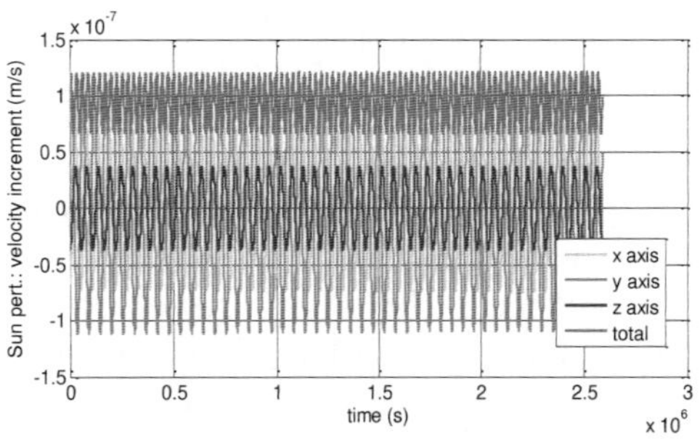

Figure 8. Sun perturbation (sim. 1)

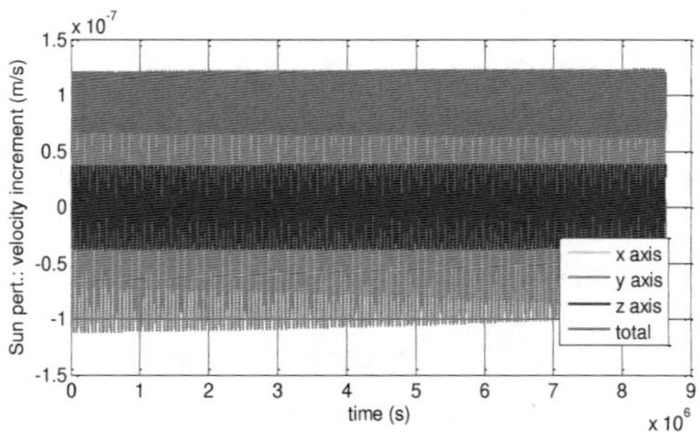

Figure 9. Sun perturbation (sim. 2)

Figures 16 and 17 show that the perturbation of the non-central field of Ida is also capable of altering the orbital plane of the spacecraft trajectory. Although the figures show small values for inclination deviations, it is clear from Figure 17 that there is a tendency of increase the deviation over time. Finally, figures 18 and 19 show the altitudes of the vehicle with respect to the asteroid Ida. Even considering almost circular trajectories a significant variation of the altitude is observed, with maximum amplitude of variation around 16 km. This is due to the irregular shape of the asteroid. It is evident in the figures that the soft landing of a spacecraft in an irregularly shaped body such as the asteroid Ida presents a great challenge [12], since the relative velocity of the vehicle and the surface of the body may vary suddenly as a function of the altitude variation, as shown in figures 18 and 19.

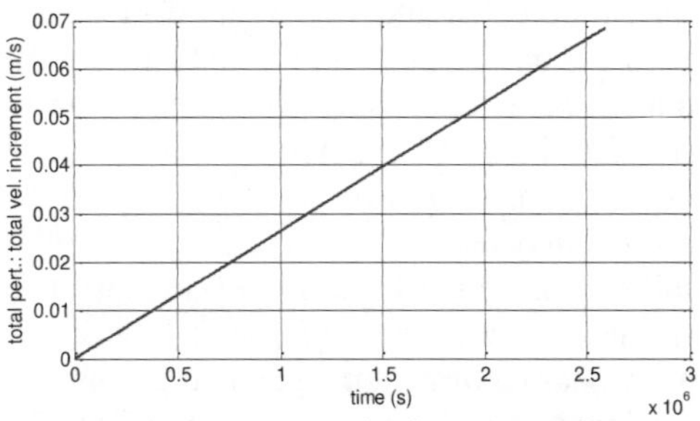

Figure 10. Perturbation Sum (sim. 1)

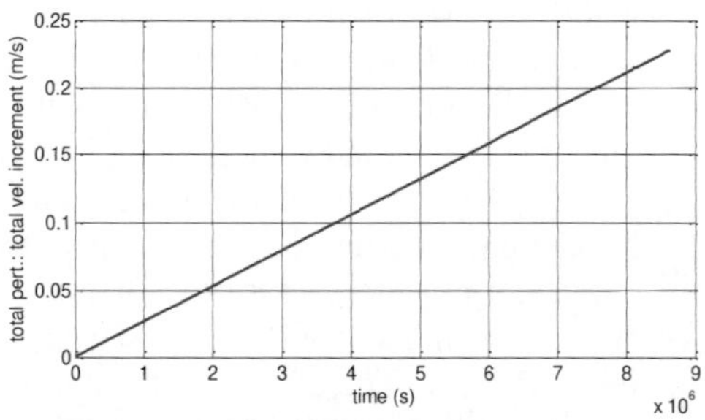

Figure 11. Perturbation Sum (sim. 2)

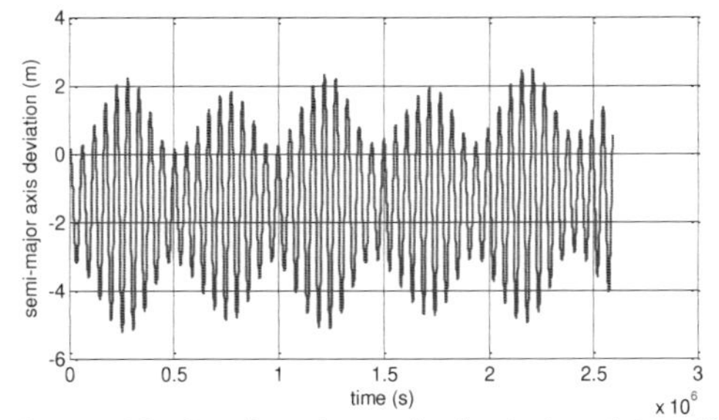

Figure 12. Semi-major axis deviation (sim. 1)

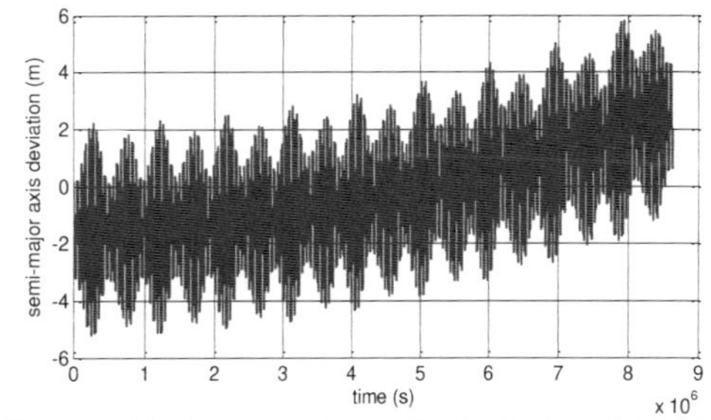

Figure 13. Semi-major axis deviation (sim. 2)

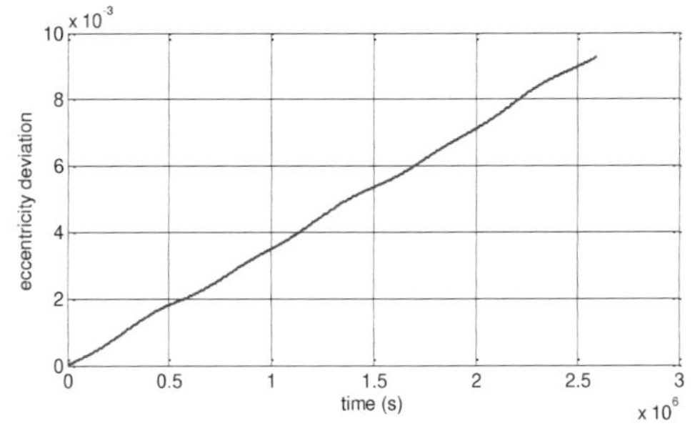

Figure 14. Eccentricity deviation (sim. 1)

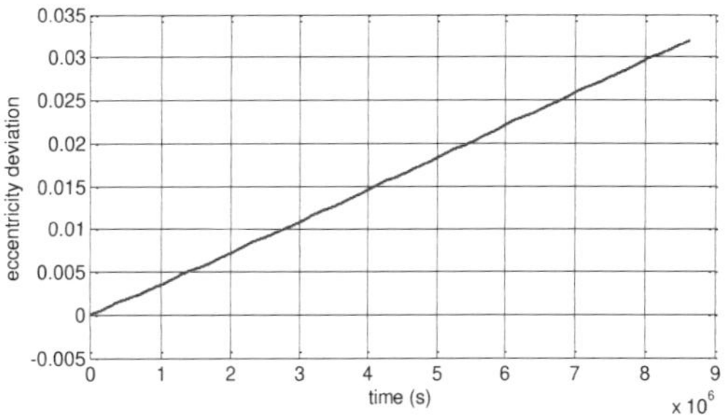

Figure 15. Eccentricity deviation (sim. 2)

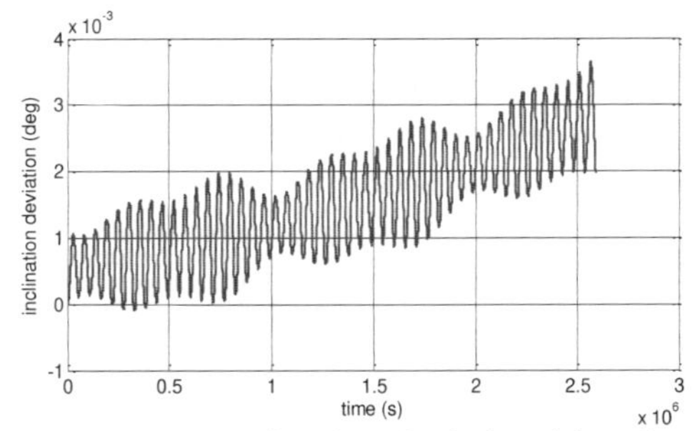

Figure 16. Inclination deviation (sim. 1)

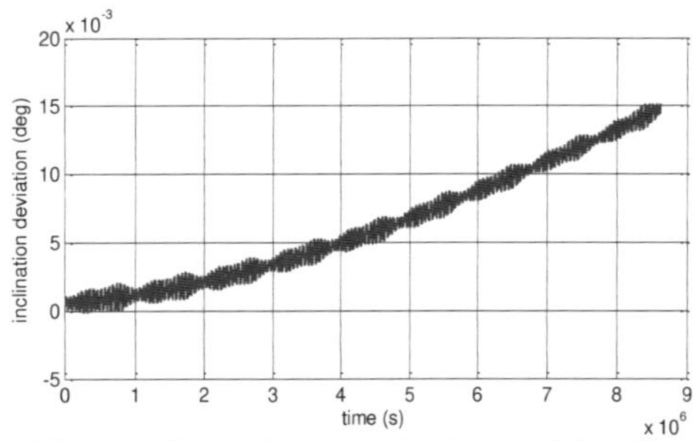

Figure 17. Inclination deviation (sim. 2)

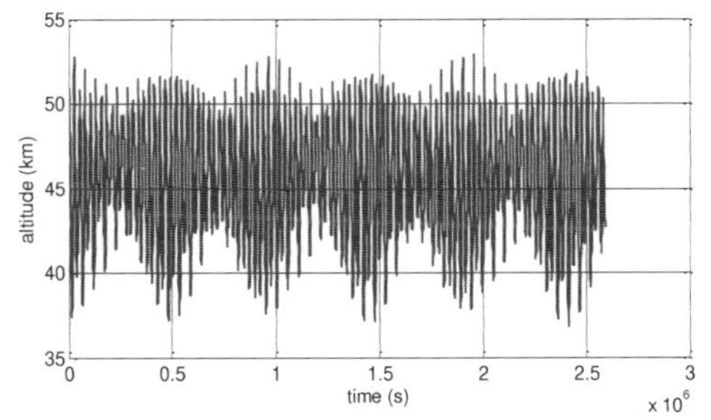

Figure 18. Altitude (sim. 1)

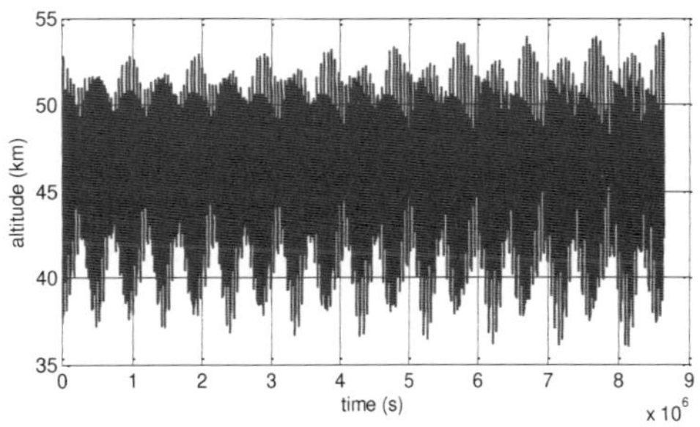

Figure 19. Altitude (sim. 2)

4. Conclusions

The simulations showed that due to the irregular shape of the asteroid the spacecraft approaches the surface even when using an almost circular initial orbit. This is a characteristic of the orbits around asteroids, in which the gravitational perturbation can oscillates expressively as function of the altitude. The main disturbance in the trajectory of the spacecraft was due to the gravitational potential of Ida. The disturbance caused by Dactyl is insignificant since the estimated mass of the natural satellite of Ida is very small, and besides that, the Dactyl orbit was not adequately determined and varies due to the gravitational perturbation of Ida. An initial orbit with an orbital period of 1.53 terrestrial days and small eccentricity was adopted, but other possible orbits should be considered. Thus, this work represents a starting point for the study of trajectories around Ida and its moon Dactyl.

References

[1] Chapman C R 1996 S-Type Asteroids, Ordinary Chondrites, and Space Weathering: The Evidence from Galileo's Fly-bys of Gaspra and Ida. *Meteoritics & Planetary Science*, v. 31(6) pp. 699–725. doi:10.1111/j.1945-5100.1996.tb02107.x

[2] Holm J; Jones J 1994 *Discovery of Ida's Moon Indicates Possible "Families" of Asteroids*. Holm J and Jones J. ed.: The Galileo Messenger v. 34. NASA. JPL 410-16-34

[3] Petit J M; Durda D D; Greenberg R; Hurford T A; Geissler P E 1997 *The Long-Term Dynamics of Dactyl's Orbit. Icarus*, v.130 (1), pp. 177–197. doi:10.1006/icar.1997.5788

[4] Mason J W 1994 Ida's new moon. *Journal of the British Astronomical Association*, v. 104 (3), pp. 108.

[5] National Aeronautics and Space Administration NASA *Images of Asteroids Ida & Dactyl* photojournal.jpl.nasa.gov/jpeg/PIA00069.jpg access 6.25.17; solarsystem.nasa.gov/galileo/

[6] Chapman C R; Klaasen K; Belton M J S; Veverka J 1994 Asteroid 243 IDA and its satellite. *Meteoritics*, v. 29, pp. 455.

[7] Stooke P 2002 *Small Body Shape Models.* EAR-A-5-DDR-STOOKE-SHAPE-MODELS-V1.0. NASA Planetary Data System

[8] Venditti F. C. F 2013 *Orbital maneuvers around irregular shaped bodies* PhD Thesis, http://urlib.net/8JMKD3MGP7W/3F7Q5U2

[9] Venditti F C F; Rocco E M; Prado A F B A 2012 *Orbital trajectories control around non-spherical bodies.* In: 9th International Conference on Mathematical Problems in Engineering, Aerospace and Sciences, Vienna, Austria. http://dx.doi.org/10.1063/1.4765622

[10] Rocco E M; Gonçalves L D; Moraes R V 2017 *Orbital maneuvers in the non-central gravitational field of Phobos perturbed by the gravitational attraction due to Mars, Sun and Deimos*. In: 27th Space Flight Mechanics Meeting, San Antonio. Proceedings of the 27th Space Flight Mechanics Meeting.

[11] Rocco E M; Gonçalves L D; Moraes R V 2017 *Orbital trajectories in the vicinity of Saturn's moons Prometheus, Epimetheus, Janus and Pandora.* In: International Conference Asteroids, Comets, Meteors 2017, Montevideo.

[12] Rocco E M 2014 *Trajetórias de aproximação para pouso no asteroide 216 Kleopatra.* In: XVII Colóquio Brasileiro de Dinâmica Orbital - CBDO 2014, Águas de Lindóia - SP - Brasil. Anais do XVII Colóquio Brasileiro de Dinâmica Orbital.

[13] Rocco E M 2008 *Perturbed orbital motion with a PID control system for the trajectory* (Águas de Lindóia: XIV Colóquio Brasileiro de Dinâmica Orbital) pp 17-21

[14] Rocco E M 2012 *Controle de trajetória com propulsão contínua para missões do tipo drag-free* (Maranhão: Congresso Nacional de Engenharia Mecância)

[15] Rocco E M 2013 Automatic correction of orbital elements using continuous thrust controlled in closed loop. *Journal of Physics Conference Series*, v. 465. doi:10.1088/1742-6596/465/1/012007

[16] Rocco E M 2015 Gravitational disturbances generated by the Sun, Phobos and Deimos in orbital maneuvers around Mars with automatic correction of the semi-major axis. *Journal of Physics Conference Series*, v. 641. doi:10.1088/1742-6596/641/1/012027

[17] Werner R. A 1994 The gravitational potential of a homogeneous polyhedron or don't cut corners. *Celestial Mechanics and Dynamical Astronomy*, v. 59, (3), pp. 253-278.

[18] Mota M L; Rocco E M 2014 *Comparação entre as abordagens impulsivas e contínua do sistema propulsivo em manobras orbitais em torno de um corpo não-esférico.* In: XXXV Iberian Latin-American Congress on Computational Methods in Engineering - CILAMCE 2014, Fortaleza - CE - Brazil. Proceedings of CILAMCE 2014.

A 2-DOF model of an elastic rocket structure excited by a follower force

Leandro F. Brejão and Reyolando Manoel L. R. da Fonseca Brasil

Federal University of ABC, UFABC, Santo Andre, Brazil

leandro.f.b.eae@gmail.com, reyolando.brasil@ufabc.edu.br

Abstract. We present a two degree of freedom model of an elastic rocket structure excited by the follower force given by the motor thrust that is supposed to be always in the direction of the tangent to the deformed shape of the device at its lower tip. The model comprises two massless rigid pinned bars, initially in vertical position, connected by rotational springs. Lumped masses and dampers are considered at the connections. The generalized coordinates are the angular displacements of the bars with respect to the vertical. We derive the equations of motion via Lagrange's equations and simulate its time evolution using Runge-Kutta 4th order time step-by-step numerical integration algorithm. Results indicate possible occurrence of stable and unstable vibrations, such as limit cycles.

1 Introduction

Launcher vehicles (popularly known as rocket) are essential devices to carry loads from the surface of the Earth to some orbit around it, whatever is the current space mission. The vehicle, like any other physical body, is not absolutely rigid, so that the structural behavior, via excitation by external loads, tends to affect the flight dynamics. It is intended, in this paper, to develop a mathematical model of an elastic space rocket structure as a Beck's column excited by a follower (or circulatory) force. This force represents the rocket motor thrust that should be always in the direction of the tangent to the structure deformed axis at the base of the vehicle. We present a simplified two degree of freedom rigid bars discrete model. Its system of two second order nonlinear ordinary differential equations of motion are derived via Lagrange's energy method, allowing for a general understanding of the main characteristics of the problem. The proposed equations consider up to third order (cubic) inertia, stiffness and forcing terms. Among other rich nonlinear dynamic behavior of this model, depending on parameters and initial conditions choices, either stable or unstable limit cycle solutions are possible. The unstable solution is, of course, an interesting simple example of flutter instability.

In the field of aerospace engineering several phenomena of interest are non-linear, as, for example, those studied by Aeroelasticity. This area of study had its development begun between 1910-1930 as presented by [1]. Aeroelastic phenomena are studied in aircraft, missiles and launch vehicles. The historical perspective on analysis and control of aeroelastic responses can be found in [2]. In particular, aeroelastic models in launch vehicles, based on articulated beams have been used by several authors as introduction to the study of this phenomenon. In [3] linear studies are developed to consider the critical load in launch vehicles. In [4] this problem is analyzed from the topological point of view of structural stability. However, in order to obtain a more adequate analysis of the structural dynamics of a launch vehicle, it is necessary to introduce non-linearities in its mathematical model. It is by means of this non-linear modeling that the determination of the amplitudes of the post-critical stationary states becomes possible, as it affirms [5] and how the work of [6] deserves attention. Another nonlinear theory associated with aeroelasticity in launch vehicles can be found in [7]. Often the study of aeroelastic phenomena in launch vehicles applies to the interaction of aerodynamic, elastic, propulsive and inertial forces acting on their structure. In particular, the events associated with the dynamic instability of harmonic and self-excited oscillatory nature occurring due to the modifying interaction between two or more distinct modes of vibration of a system and the introduction of external energy into that system by means of An agent or dynamic flow (force), according to the definitions found in [8], [9] and [10]. In the present study, this energy introduction is associated with the axial circulatory load at the base of the launch vehicle because of its propulsion. In [11] an analytical solution of the fourth-order partial differential equation of the Beck column is also presented. Some thought will also be given in the future to possible control strategies of the resulting vibrations, a crucial problem when real rocket motions are considered. In future further work, we will develop the exact analytical solution to the fourth order partial differential equation governing the motion of an elastic space rocket structure under follower force excitation.

2 The structural physical model and modeling hypothesis

Fig. 1 is the simplified physical model of the structure of a launcher vehicle. It is constructed of two rigid massless bars $\overline{AB}$ and $\overline{BC}$, L_1 and L_2 long, respectively, pinned to nodes A and B. Displacements are restricted at point A. We consider lumped masses M_1, M_2 and M_3 attached to nodes A and B where torsional springs k_1 e k_2 provide elastic restoring forces. Then, viscous dampers of damping constants $\hat{\mu}_1$ and $\hat{\mu}_2$ are added to the joints of the system.

1. It is adopted $L_1 = L_2 = L$.
2. The bars are rigid and massless.
3. Lumped masses M_1, M_2, and M_3 represent the actual masses of half the bars connected to that point. If $2m$ is the mass of each bar, $M_1 = M_3 = m$ and $M_2 = 2m$.
4. The stiffness constants of the torsion springs of the system are considered to represent the stiffness of the constituent material of the bars. In particular, the relation $k_1 = k_2$ is adopted for this model. These springs are assumed to be linear, that is, the restoring moments they apply to the structure (elastic moments) are proportional to the angular displacements comprised between the longitudinal axis of each bar and the local vertical.
5. The viscous damping constants of the shock absorbers of the system are considered to represent the dissipative tendency of energy by the structure of the vehicle given its deformation. In particular, the relation $\hat{\mu}_1 = \hat{\mu}_2 = \hat{\mu}$ is adopted for this model. The dampers are assumed to be linear, i.e. the dissipation moments they apply to the structure are proportional to the first derivatives of the angular displacements comprised between the longitudinal axis of each bar and the local vertical.
6. Motions are restricted to the Axy plan.
7. Initially, only self-weight forces act. This is the fundamental static equilibrium configuration of the system, representing the vehicle at rest in its launch platform.

For launch vehicles, it is common to consider a fixed reference coordinate system on its nose (the end of the warhead), with which one can define the vehicle's static margin. In this study, the coordinate system is fixed at A, as shown in Fig. 1. This system is non-inertial for a solo observer but is assumed to be inertial in the frame of the vehicle and as soon as it travels with it. For this reason, an articulated support in A and elastic link between this support and the vehicle is assumed.

3 Physical loading model

Let F be a follower (circulatory) non-conservative force applied to C, in the direction of bar BC. This force models the rocket's thrust force due to combustion gases expansion at the motors in the basis of the vehicle. We do not consider, in this model, its dependence on time.

The action of force F applied to C excites the system to depart from its fundamental equilibrium position. The problem is now similar to an excited inverted double-pendulum with elastic properties. The equations of motion could be derived via Newton's second law vector approach, but this method is quite cumbersome in this case. Thus, a Lagrangian scalar energy scheme is preferable. Our generalized coordinates are angular displacements θ_1 and θ_2 of bars $\overline{AB}$ and $\overline{BC}$, computed from their original vertical equilibrium positions. They are, of course, implicitly time dependent, that is, $\theta_1 = \theta_1(t)$ and $\theta_2 = \theta_2(t)$. We denote $\theta_1(t) \equiv q_1(t)$ and $\theta_2(t) \equiv q_2(t)$. Nonzero values represent the vehicle in flight conditions as represented in Fig, 1.

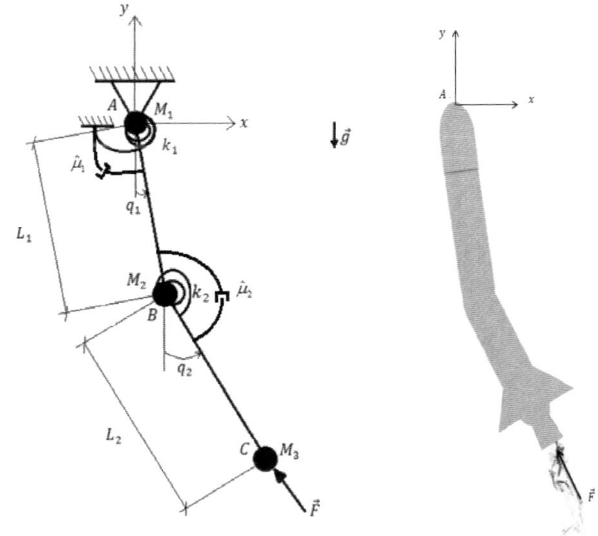

Fig.1: (left) System motion under follower force **F**; (right) rocket appearance under follower force **F.**

4 Mathematical model

4.1 Kinematics

4.1.1 Position vectors of the lumped masses

$$r_1 = 0 \tag{1}$$

$$r_2 = L\left(\sin q_1 \hat{i} - \cos q_1 \hat{j}\right) \tag{2}$$

$$r_3 = L\left[\left(\sin q_1 + \sin q_2\right)\hat{i} - \left(\cos q_1 + \cos q_2\right)\hat{j}\right] \tag{3}$$

4.1.2 Velocity vectors of the lumped masses

$$r_1 = 0 \tag{4}$$

$$r_2 = L\left(q_1 \cos q_1 \hat{i} + q_1 \sin q_1 \hat{j}\right) \tag{5}$$

$$r_3 = L\left[\left(q_1 \cos q_1 + q_2 \cos q_2\right)\hat{i} + \left(q_1 \sin q_1 + q_2 \sin q_2\right)\hat{j}\right] \tag{6}$$

4.1.3 Approximations

We adopt a third order truncated polynomial McLaurin approximations to the sinusoidal functions.

$$\cos q_i \cong 1 - \frac{q_i^2}{2}, \qquad \sin q_i \cong q_i - \frac{q_i^3}{6} \tag{7}$$

4.2 Energy computation
In the following equations, we neglect terms of higher order than third.

4.2.1 Kinetic energy

$$T \cong \frac{1}{2} mL^2 \left[3q_1{}^2 + q_2{}^2 + 2\left(q_1 q_2 - q_1 q_2 \frac{q_1{}^2}{2} - q_1 q_2 \frac{q_2{}^2}{2} + q_1 q_2 q_1 q_2 \right) \right] \tag{8}$$

4.2.2 Total potential energy

$$V = U - W = \frac{1}{2} k \left(2q_1{}^2 + q_2{}^2 - 2q_1\, q_2 \right) = \frac{1}{2} mgL \left(3q_1{}^2 + q_2{}^2 \right) \tag{9}$$

where U is the strain energy of the springs and W is the work of the conservative forces (masses weight).

4.3 Derivation of the equations of motion
Next, we apply Euler-Lagrange equations to the Lagrangian functional $\mathcal{L} = T - V$:

$$\frac{\mathrm{d}}{\mathrm{d}t}\left(\frac{\partial}{\partial q_i} \mathcal{L} \right) - \frac{\partial}{\partial q_i} \mathcal{L} = F_i^{nc} = \sum_{j=1}^{N} f_j^{nc} \bullet \frac{\partial}{\partial q_i} r_j \tag{10}$$

where the non-conservative generalized forces are in the left-hand term. The detailed mathematical deductions of this study can be found in [5], [12] and [13].

4.4 Generalized non-conservative forces
The generalized non-conservative forces, due to the follower force **F** are:

$$f_1^{nc} = f_2^{nc} = 0, \quad f_3^{nc} = F\left(-\sin q_2 \hat{i} + \cos q_2 \hat{j} \right) \tag{11}$$

$$F_1^{nc} \cong -FL\left(-q_1 + q_2 + \frac{q_1{}^3}{6} - \frac{q_2{}^3}{6} + \frac{q_1 q_2{}^2}{2} - \frac{q_2 q_1{}^2}{2} \right), \quad F_2^{nc} = 0 \tag{12}$$

where F is the scalar value of the follower force, considered time independent.

5 Equations of motion

Although gravity has been considered in the process of mathematical modeling of the problem, present in the working terms of the conservative forces acting on the vehicle (weight forces), from now on, the gravitational effect on the dynamics of the system is ignored due to the comparison established with the Propulsion force. This situation models the vehicle's flight at high altitudes, in which the Earth's gravitational acceleration can be neglected for the purpose of simplifying the problem. In fact, as we know, the Euclidean metric norm of acceleration of Earth's gravity decays with the square of the distance between the center of mass of the body under study (in this case the vehicle) and the center of mass of the Earth. It is an interesting fact, though not impressive, that the acceleration of gravity contributes to the dynamic stability of the system. In order to analyze the dynamics of the vehicle subjected to the loading from the propulsion, so that the effects of this one became more evident for study, the rigidity associated to the work of the conservative forces was suppressed.

IOP Conf. Series: Journal of Physics: Conf. Series **911** (2017) 012020 doi:10.1088/1742-6596/911/1/012020

$$\left[\begin{bmatrix}3 & 1\\ 1 & 1\end{bmatrix}+\begin{bmatrix}0 & -\frac{1}{2}(q_1-q_2)^2\\ -\frac{1}{2}(q_1-q_2)^2 & 0\end{bmatrix}\right]\begin{Bmatrix}\ddot{q}_1(t)\\ \ddot{q}_2(t)\end{Bmatrix}+$$
$$\left[\begin{bmatrix}0 & \dot{q}_2(q_1-q_2)\\ -\dot{q}_1(q_1-q_2) & 0\end{bmatrix}+\mu\begin{bmatrix}2 & -1\\ -1 & 1\end{bmatrix}\right]\begin{Bmatrix}\dot{q}_1(t)\\ \dot{q}_2(t)\end{Bmatrix}+c\begin{bmatrix}2 & -1\\ -1 & 1\end{bmatrix}\begin{Bmatrix}q_1(t)\\ q_2(t)\end{Bmatrix}= \tag{13}$$
$$-\lambda c\begin{Bmatrix}-q_1+q_2+\frac{q_1^3}{6}-\frac{q_2^3}{6}+\frac{q_1q_2^2}{2}-\frac{q_2q_1^2}{2}\\ 0\end{Bmatrix}$$

Where

$$c=\frac{k}{mL^2},\quad \lambda=\frac{FL}{k},\quad \mu=\frac{\hat{\mu}}{mL^2} \tag{14}$$

6 Results

Next, computational simulations are presented, from the parameterized model of Eq. (13), for a case in which the launch vehicle model is stable, a case of limit cycle, and a case of flutter-type intactness. In the scope of this study, the notion of stability is associated with the limited output relationship - amplitude of the response, which is the generalized displacements of the model - for limited input, which is the follower force, of constant intensity in time. The simulations were obtained for the initial conditions $\theta_1(t) = q_1(t) = 0.01$ rad, $\theta_2(t) = q_2(t) = 0.01$, $\dot{\theta}_1(t) = \dot{q}_1(t) = 0$ rad and $\dot{\theta}_2(t) = \dot{q}_2(t) = 0$ rad. These conditions represent the axis of symmetry of the rocket slightly misaligned with the vertical reference direction. It is important to emphasize that instability in the example of post-critical steady regime of the limit cycle type is characteristic of the physical phenomenon and not of the integrator the 4th order Runge-Kutta. Several integration steps were tested in the simulations and the same dynamic temporal behavior of the physical system was observed.

6.1 A case of asymptotic *stability*

As an example of a case of stability of the system modeled under load action at its base, the arrangement m = 1 kg, L = 1 m, k = 1 Nm/rad, $\hat{\mu}$ = 0.1 kg/s kg/s and P = 1 N is considered. The results in the configuration are shown Fig. 2:

6.2 A case of post-critical stationary regime type limit cycle

As an example of a case of limit cycle of the system modeled under load action at its base, the arrangement $m = 1$ kg, $L = 1$ m, $k = 1$ Nm/rad, $\hat{\mu} = 2$ kg/s and $P = 4$ N is considered. The results in the configuration are shown Fig. 3:

7 Conclusions and future research

We presented a two degree of freedom lumped parameters model of an elastic rocket structure excited by the follower force given by the motor thrust that is supposed to be always in the direction of the tangent to the deformed shape of the device at its lower tip. We derived the equations of motion via Lagrange's equations and simulate its time evolution using Runge-Kutta 4th order time step-by-step numerical integration algorithm. Results indicate possible occurrence of stable and unstable vibrations, such as limit cycles. research on flutter instability is ongoing.

Acknowledgements

The authors acknowledge support by CNPq, CAPES and FAPESP, all Brazilian research funding agencies.

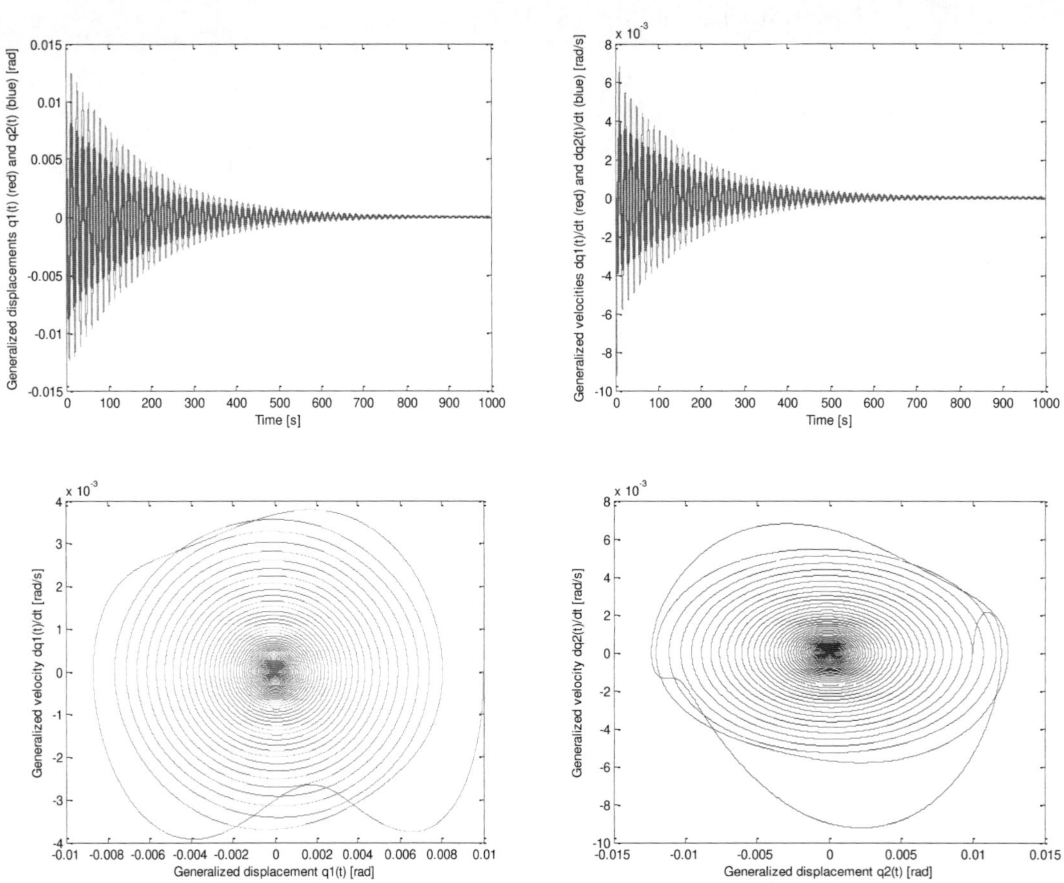

Fig. 2: (above and left) Generalized displacements in stable condition; (above and right) rates of generalized displacements in stable condition; (below and left) phase plane of generalized displacement q_1 in stable condition; (below and right) phase plane of generalized displacement q_2 in stable condition.

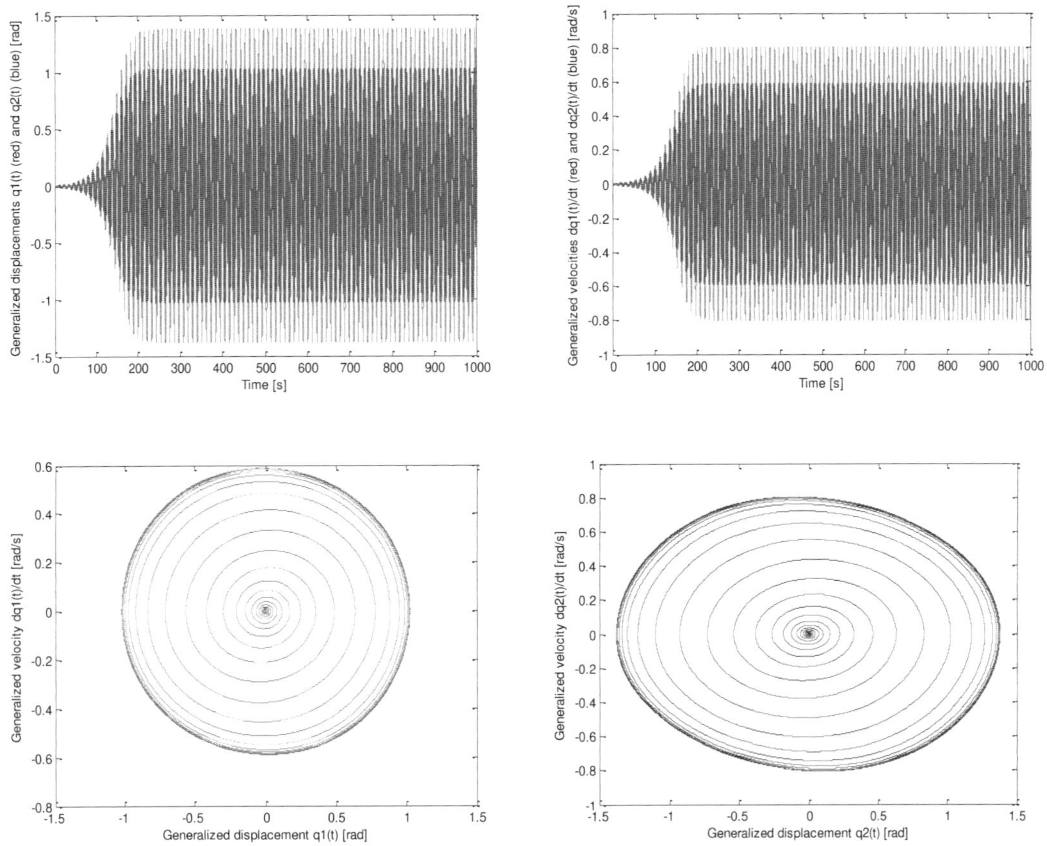

Fig. 3: (above and left) generalized displacements in limit cycle case; (above and right) rates of generalized displacements in limit cycle case; (below and left) phase plane of generalized displacement q_1 in limit cycle case; (below and right) phase plane of generalized displacement q_2 in limit cycle case

References

[1] D. H. Hodges,G. A. Pierce, Introduction to structural dynamics and aeroelasticity, New York Cambridge University Press (2002).

[2] V. Mukhopadhyay, Historical perspective on analysis and control of aeroelastic responses, Journal of Guidance, Control and Dynamics (2003) 673–684.

[3] G. Herrman, I. Jong, On the destabilizing effect of daming in nonconservative elastic systems, J. Appl. Mech (1965) .

[4] J. M. T. Thompson, Instabilities and catastrophes in science and engineering, John Wiley, Chichester (1982).

[5] C. E. N. Mazzilli, Dinâmica não linear e estabilidade: uma formulação para sistemas submetidos a excitação dedsuporte ou carregamentos não conservativos. Tese de Livre Docência, Escola Politécnica da USP, São Paulo (1988).

[6] J. Roorda, S. Nemat-Nasser, An energy method for analysis of nonlinear, nonconservative systems, AIAA J. (1967) 1262-1268.

[7] J. Dugundji, Nonlinear problems of aeroelasticity computational nonlinear mechanics in aerospace engineering, In AtluriSN, editor. Progressin astronautics andaeronautics. Washington,DC:AIAA (1992) 127.

[8] R. L. Bisplinghoff, H. Ashley, and R. Halfman, Aeroelasticity, Addison Wesley (1955).

[9] E. H. Dowell,etal. A Modern Coursein Aeroelasticity., 4. ed., Kluwer Academic (2005).

[10] Y. C. Fung, An Introduction to the Theory of Aeroelasticity, Wiley (1955).

IOP Conf. Series: Journal of Physics: Conf. Series **911** (2017) 012020 doi:10.1088/1742-6596/911/1/012020

[11] S. P. Timoshenko, Theory of elastic Stability 2nd Ed., Dover, Mineola (2009).

[12] R. M. L. R. F. Brasil, O Fenômeno de Localização de Modos em Dinâmica e Estabilidade de Estruturas Modulares de Comportamento Linear ou Não-Linear, Tese de Livre Docência, Escola Politécnica da USP, São Paulo (1996).

[13] L.F. Brejão, R.M.L.R.F. Brasil, Modeling a Rocket Elastic Structure as a Beck's Column Under Follower Force. In: Proceedings of CILAMCE 2016. Brasília (2016).

IOP Conf. Series: Journal of Physics: Conf. Series **911** (2017) 012021 doi:10.1088/1742-6596/911/1/012021

Particle-in-cell numerical simulations of a cylindrical Hall thruster with permanent magnets

Rodrigo A. Miranda[1,2], Alexandre A. Martins[2] and José L. Ferreira[2]

[1] UnB-Gama Campus, University of Brasília (UnB), Brasília DF 70910-900, Brazil
[2] Plasma Physics Laboratory, Institute of Physics, University of Brasília (UnB), Brasília DF 70910-900, Brazil

E-mail: rmiracer@unb.br

Abstract. The cylindrical Hall thruster (CHT) is a propulsion device that offers high propellant utilization and performance at smaller dimensions and lower power levels than traditional Hall thrusters. In this paper we present first results of a numerical model of a CHT. This model solves particle and field dynamics self-consistently using a particle-in-cell approach. We describe a number of techniques applied to reduce the execution time of the numerical simulations. The specific impulse and thrust computed from our simulations are in agreement with laboratory experiments. This simplified model will allow for a detailed analysis of different thruster operational parameters and obtain an optimal configuration to be implemented at the Plasma Physics Laboratory at the University of Brasília.

1. Introduction

Hall thrusters (HTs) are electromagnetic propulsion devices that apply a cross-field discharge in an annular channel to generate a plasma [1]. A radial magnetic field is responsible for electron trapping in the channel, forming the electron Hall current and a virtual cathode that helps ionize the neutral gas. An electric field is applied in the axial direction, between the anode and the virtual cathode, and is responsible for ion acceleration. Therefore, the strength of the magnetic field must be chosen to keep only the ions unmagnetized. HTs can achieve higher thrust levels compared to ion thrusters, and are simpler to construct. However, their scalability to smaller dimensions compatible for micro and nano-satellites present several challenges. For example, since the surface-to volume ratio in the annular channel increases, the collisions between particles and walls increase as well, leading to heating and higher erosion rates. Moreover, the reduced thruster dimensions makes the magnetic circuit difficult to maintain. The scaling down of HTs to various power levels has been the subject of several studies [2, 3, 4, 5].

The cylindrical Hall thruster (CHT) is an alternative concept first developed at the Princeton Plasma Physics Laboratory [6]. It consists of a cylindrical region and a short annular channel with variable depth, and presents an increased volume-to-surface ratio in comparison to compact HTs, potentially reducing wall erosion. The CHT presents higher propellant utilization and comparable performance with conventional HTs [7]. The magnetic field in this thruster can be set by using electromagnets or permanent magnets [8].

The CHT has been characterized through laboratory experiments [6, 8, 9, 10] and numerical simulations [11, 7, 12]. Several strategies have been applied to model the CHT numerically. For example, Ref. [11] used particle-tracing to study the cross-field diffusion of electrons in a typical

IOP Conf. Series: Journal of Physics: Conf. Series **911** (2017) 012021 doi:10.1088/1742-6596/911/1/012021

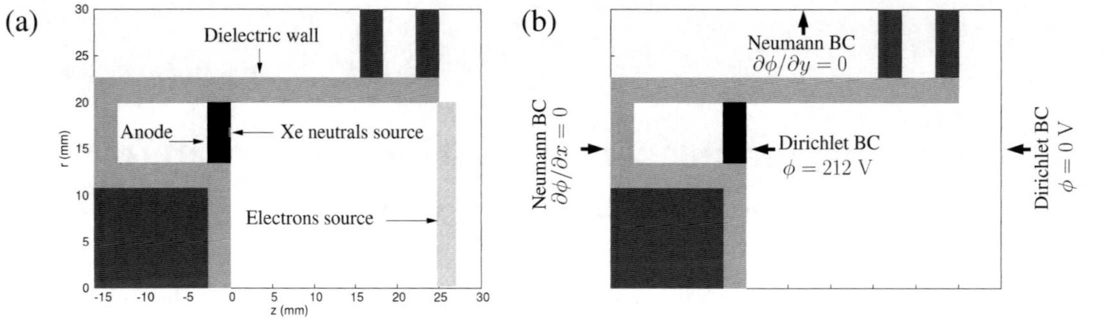

Figure 1. The simulation domain representing the upper-half cross-section of a cylindrical Hall thruster, showing (a) particle sources and boundaries, and (b) boundary conditions of the electric potential ϕ. Blue represents the structure supporting the dielectric wall.

magnetic field configuration of a two-dimensional (2D) model of a CHT. They showed that the electrons are most concentrated near the annular region of the CHT. Ref. [7] applied an hybrid model consisting of a fluid description of electrons and particle description of neutrals and ions to a 2D model of a CHT. Their results show that the ionization occurs in the annular region, and that the doubly charged ions play an important role on the level of propellant efficiency. A three-dimensional particle-in-cell (PIC) model of a CHT was proposed and studied by Ref. [12]. They were able to reproduce asymmetric features of the Hall current known as "spokes" commonly observed in laboratory experiments [13], and found that the depletion of neutral gas leading to azimuthal asymmetry of the discharge is a possible cause of these features.

The Plasma Physics Laboratory at the University of Brasília has been developing Hall thrusters with permanent magnets for the Brazilian Space Program since 2004 [14]. The team is currently investigating alternatives to conventional HTs, such as the CHT. The reduced dimensions of the CHT operating at lower power levels makes it an interesting option of electric propulsion device for future Brazilian space missions. A numerical model is in development for testing operational parameters and performance for a future CHT prototype to be constructed at the laboratory.

In this paper we present the first results of a 2D model of a CHT using PIC code. This paper is organized as follows. Section II introduces the numerical model of a CHT. Our numerical results are presented in Section III. The conclusions are indicated in Section IV.

2. Numerical model

Following previous works [11, 9, 7] we adopt cylindrical coordinates and neglect variations in the azimuthal direction to define a two-dimensional simulation domain representing the upper-half of a cross-section of the CHT. The 2D axisymmetric model allows to reduce the time needed to perform computations. The geometry of the simulation domain, including particle sources and boundary conditions of particles and fields, is shown in Figure 1. We dispose the axial direction z as the horizontal axis, and the radial direction r as the vertical direction. The size of the simulation domain is $[-16, 30] \times [0, 30]$ mm.

2.1. Field solver

The thruster's magnetic field $\mathbf{B}$ is assumed to be static. The magnetic field generated by permanent magnet rings is modelled and computed using the Comsol software [15]. The resulting vector field is depicted in Figure 2(a). Fig. 2(b) shows the modulus of the magnetic field $|\mathbf{B}|$ as

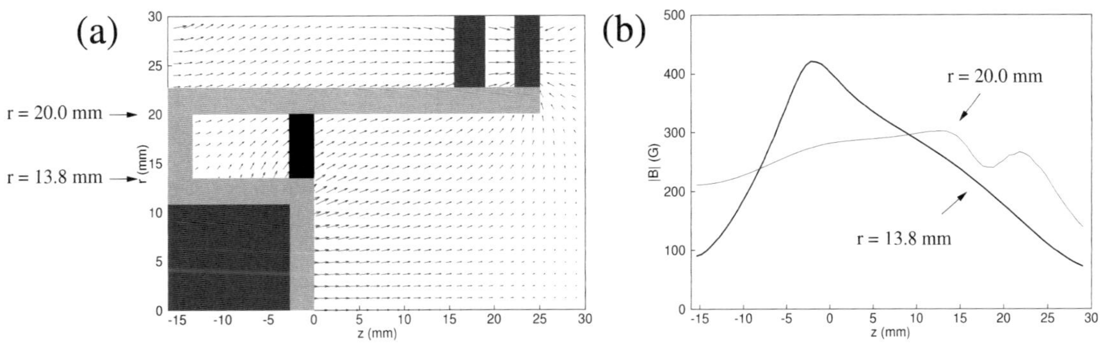

Figure 2. (a) The magnetic vector field $\mathbf{B}(z, r)$ in the simulation domain. (b) The modulus of the magnetic field, $|\mathbf{B}|$, as a function of the axial distance z, at r = 13.8 mm (thick line), and r = 20 mm (thin line).

a function of the axial distance, at $r = 13.8$ mm and at $r = 20$ mm, which are the position of the inner dielectric surface at the core and the position of the outer dielectric surface, respectively. The $|\mathbf{B}|$ profiles are in agreement with laboratory experiments [6, 10].

The electric potential ϕ is solved using a self-consistent algorithm described in the next section. We set different boundary conditions for the electric potential ϕ as follows. Neumann boundary conditions are set at the $z = -16$ mm and $r = 30$ mm boundaries. At $z = 30$ mm we set a Dirichlet boundary condition with $\phi = 0$ V. The anode is represented by a region in which $\phi = 212$ V using Dirichlet boundary conditions. These boundary conditions are depicted in Fig. 1(b).

2.2. Particle solver

We model charged and neutral particles using a PIC description. In this approach, the positions and velocities of charged particles (i.e., electrons and xenon ions) are computed from the equations of motion due to the presence of electric and magnetic fields. The charge density is then discretized to a spatial grid by a weighting process, which allows to solve the Poisson equation to obtain new values of the electric potential and the electric field. These fields are then inserted into the equations of motion of the particles, and the cycle is restarted. Particle simulations are performed using the VSim software [16]. The fields are solved in a spatial grid of size 150×255.

Representing all plasma particles in the simulations of a CHT renders the computational effort practically impossible to reach. A common approach to reduce the computation time in PIC simulations is to simulate superparticles, which are particles that represent a "cloud" or a group of electrons or ions. The position of the superparticle is given by the center of mass of the cloud, and its velocity is given by the mean velocity of the cloud [17]. Therefore, the number of particles present in the simulation is reduced as well as the execution time. We use the superparticle approach to represent electrons, Xe^+ and xenon neutrals (Xe). Neutral superparticles represent 10^7 real Xe particles, whereas electrons and Xe ions superparticles represent 10^5 particles. In the following we will refer to the superparticles as simply "particles".

We model Xe neutrals using superparticles with no charge. In laboratory experiments, the xenon gas is injected through small holes in the CHT anode. To mimic this, the neutral particles enter the simulation domain through a narrow region centered in front of the anode (see Fig. 1(a)). Their initial velocity is taken from a Maxwellian distribution with temperature of 560 K, and the flow rate is set to 4 sccm. Neutral particles that hit the dielectric and anode walls are

Parameter	Scaling law
Length	$L^* = \zeta L$
Electric potential	$\phi^* = \phi$
Magnetic field	$\mathbf{B}^* = \zeta^{-1}\mathbf{B}$
Mass flow rate	$\dot{m}^* = \zeta \dot{m}$
Current	$I^* = \zeta I$
Electric field	$\mathbf{E}^* = \zeta^{-1}\mathbf{E}$
Particle number density	$n^* = \zeta^{-1} n$
Temperature	$T^* = T$

Table 1. The scaling laws applied. An asterisk indicates the scaled variable.

reflected with a velocity taken from a thermal distribution. Particles that leave the simulation domain are removed.

Electrons are inserted in the simulation through a narrow region located in front of of the exit region as shown in Fig. 1(b). Trying to represent a cathode as the source of electrons would break the cylindrical symmetry of our model. For this reason we simplify the geometry and insert the electrons near the exit region of the CHT with an initial velocity taken from a Maxwellian distribution with temperature of 10 eV. This approach has been used in previous simulations of CHTs [11, 9]. Electrons leaving the simulation domain are removed.

2.3. Particle interactions

The interaction between electrons and Xe neutrals are solved by the VSim software using a MonteCarlo collision (MCC) technique. We implement (Xe, e^-) ionization collisions, (Xe, e^-) elastic scattering, (Xe, e^-) excitation collisions, (Xe, Xe^+) elastic scattering, and (Xe, Xe^+) momentum exchange. When an ionization collision occurs, ions and secondary electrons are created at the position of the incident electron. The energy of the created ion is determined from the temperature of the corresponding neutral particle. The energy of the incident electron decreases due to the ionization threshold energy and the energy of the secondary electron.

2.4. Scaling laws

In order to speed up the execution of the PIC code further, we apply a self-similar scaling ζ following Ref. [18]. This scaling of the CHT geometry is applied to reduce the channel dimensions linearly while at the same time keeping the main plasma parameters constant. A set of scaling laws is applied to the main variables such as the magnetic field, the electric field, and the mass flow rate. A summary of the scaling laws applied is given in Table 1. The simulation execution time is then reduced, and the original system can be recovered by applying the inverse scaling law. Here we use $\zeta = 0.1$.

3. Simulation results

We start our simulations by setting the electron source disabled and the Xe source enabled. By doing so, we allow the Xe particles to enter the simulation domain and fill the annular and cylindrical regions. Since the equations of fields and charged particle motion are not solved in this initial stage, the time step is set to $\Delta t = 10^{-8}$ s, which is larger than the time step needed in the presence of charged particles. Fig. 3(a) shows the time series of the Xe neutral density averaged in the region given by $0 < z < 25$ mm and $0 < r < 20$ mm, i.e., the cylindrical region. The average Xe density quickly reaches a steady state. Fig. 3(b) shows a snapshot of the resulting Xe neutral density at $t = 0.01$ s. There is a maximum of neutral Xe density at the

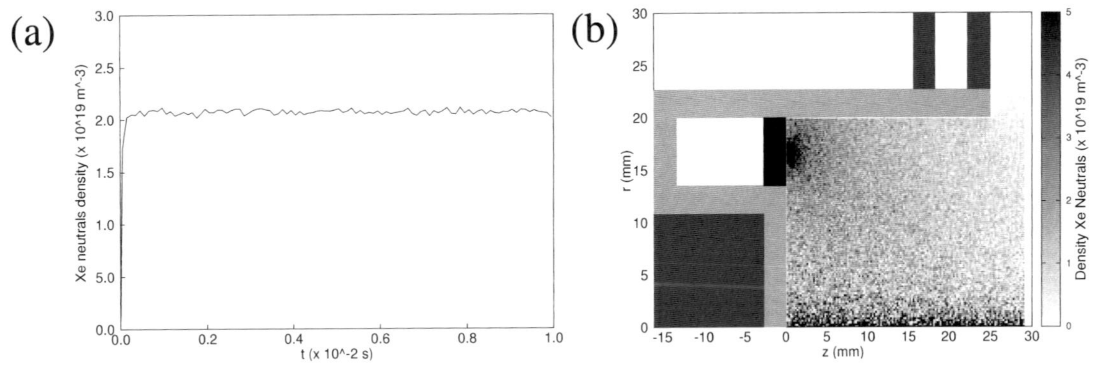

Figure 3. (a) Time series of the average of the Xe neutral density, represented by the gray color. (b) The Xe neutral density at $t = 0.01$ s.

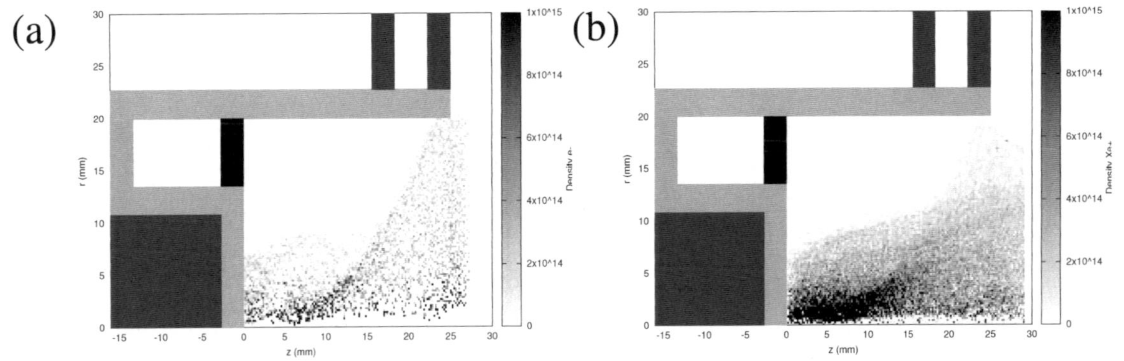

Figure 4. (a) The electron density, and (b) the ion Xe^+ density, at $t = 6.75 \times 10^{-7}$ s, represented using the gray color.

source region in front of the anode, and a maximum along the z axis due to converged flux near the cylindrical axis of symmetry.

After $t = t_0 = 0.01$ s electrons are introduced by enabling the electron source (see Fig. 1(a)). The time step is reduced to $\Delta t = 10^{-12}$ s to account for the plasma time scales such as the electron plasma frequency. In the following we refer to the simulation time as the time after t_0, i.e., $t - t_0$. Fig. 4(a) shows the electron density at $t = 6.75 \times 10^{-7}$ s. The electrons move toward the anode, and are confined near the center of the cylindrical region due to a combined effect of the electric potential and the magnetic field gradient. The confinement of electrons is clearly seen as a maximum of the electron density in Fig. 4(a).

The electrons ionize the neutral gas producing Xenon ions. Fig. 4(b) shows the density of Xe^+ at $t = 6.75 \times 10^{-7}$ s. Since the ions are unmagnetized they are repelled toward the exit region due to the electric potential. A maximum value of the Xe^+ density is observed at the center of the cylindrical region near the dielectric wall. From Fig. 4 we can conclude that the ionization is enhanced in this region.

Since the velocity of ions greatly exceeds the velocity of Xe neutrals in the exhaust region of conventional Hall thrusters, the thrust T can be written as [1]

$$T = \dot{m}_i \langle v_i \rangle \tag{1}$$

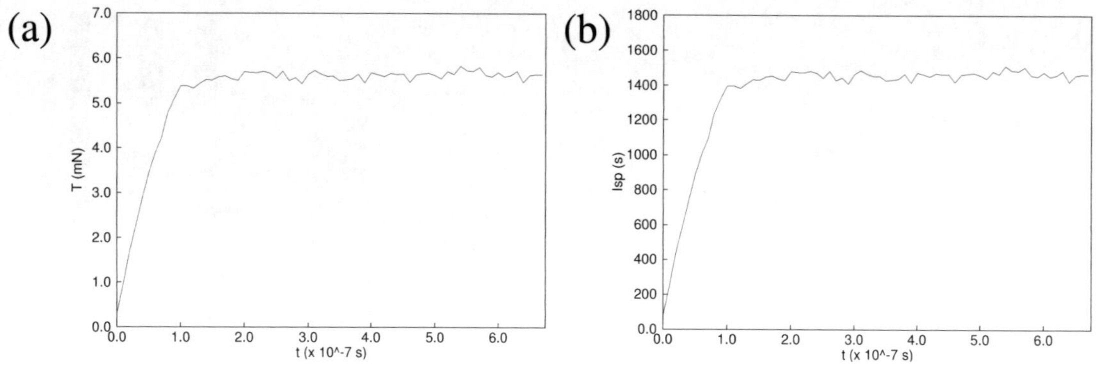

Figure 5. (a) Time series of thrust T. (b) Time series of the specific impulse I_{sp}.

where $\dot{m}$ is the ion mass flow and $\langle v_i \rangle$ is the average ion velocity. Equation 1 can be computed from numerical simulations using the discretized form

$$T = N_p \frac{m_i}{\Delta t} \langle v_i \rangle \tag{2}$$

where N_p is the number of particles represented by a superparticle, and Δt is the time step. The specific impulse I_{sp} is related to the thrust by

$$I_{sp} = \frac{T}{\dot{m} g} \tag{3}$$

where $\dot{m}$ is the Xe neutral mass flow rate, and $g = 9.8$ m/s^2 is the acceleration of gravity.

Figure 5 shows the time series of T and I_{sp} computed using Eqs. (2) and (3), respectively, from $t = 0.0$ s (i.e., when the electron source is enabled) to $t = 6.75 \times 10^{-7}$ s. There is an initial transient period until $t = 1 \times 10^{-7}$ s, after which both T and I_{sp} reach a steady state. The average value of T and I_{sp} computed during the interval $[2 \times 10^{-7}, 6.75 \times 10^{-7}]$ s gives $\langle T \rangle = 5.6 \pm 0.9$ mN, and $\langle I_{sp} \rangle = 1460.4 \pm 22.0$ s, which are close to the values reported in laboratory experiments [10].

4. Conclusions

In this paper we described a model of a cylindrical Hall thruster to be implemented at the Laboratory of Plasmas at the University of Brasília. The model describes the interaction of neutral and charged particles with electric and magnetic fields using a PIC technique. We applied several methods to reduce the execution time and computational effort, for example, by using superparticles [17] and geometrical scaling laws [18]. We presented results on the Xe neutrals, Xe^+ ions and electron density. We compute the thrust and the specific impulse, and our results are in agreement with laboratory experiments. The simplified model presented here will allow to explore in detail the parameter space of the CHT to find an optimal configuration.

The electron and Xe^+ densities presented in this paper show that the ionization is higher in a central region near the dielectric wall. The electron mobility in the direction perpendicular to magnetic field lines can be affected by the reduction of dimensions in 2D models. For this reason, several authors include an additional electron transport mechanism to account for this effect in numerical simulations of Hall thrusters [19, 20, 21, 22]. By considering an enhanced electron transport, the ionization region is higher near the anode, where the magnetic field lines accumulate [11, 7]. However, the origin of this anomalous cross-field transport is not well understood [23, 20, 24]. Moreover, there is no general agreement on how this effect should be included in numerical simulations [20, 24, 22]. The inclusion of a term representing anomalous electron diffusion in our model, for example, Bohm diffusion [11] will be the focus of future work.

5. Acknowledgements

The authors acknowledge support from the UNIESPAÇO program of the Brazilian Space Agency (AEB), the National Council of Technological and Scientific Development (CNPq), FAPDF, IF/UnB, DPP/UnB and CAPES. R.A.M. acknowledges support from FAPDF under grant 0193.000984/2015. A.A.M. gratefully acknowledges AEB, the National Institute for Space Research (INPE) and CNPq for the attribution of a post-doc scholarship for the development of electric propulsion systems to be used in satellites.

References

[1] Goebel D M and Katz, I 2008 *Fundamentals of Electric Propulsion: Ion and Hall Thrusters* (JPL Space Sci. Tech. Series).
[2] Kim V, Kozlov V, Kazurenko A, Popov G, Skrylnikov A, Clauss C, Day M and Sankovic J 1998 *34th Joint Propulsion Conf.* (Cleveland) AIAA 98-3335.
[3] Beal B E, Gallimore, A D 2005 *Phys. Plasmas* **12** 123503.
[4] Ding Y J, Yu D R, Kia D C, Yan G J and Li H 2011 *Contrib. Plasma Phys.* **51** 68.
[5] Ahedo E and Gallardo J M 2003 *28th Intern. Electric Propulsion Conf.* (Tolulouse) IEPC 2003-104.
[6] Raitses Y and Fisch N J 2001 *Phys. Plasmas* **8** 5 2579.
[7] Garrigues L, Hagelaar G J M, Boeuf J P, Raitses Y, Smirnov A and Fisch N J 2008. *IEEE Trans. Plasma Sci.* **35** 5 2008.
[8] Raitses Y, Merino E, Fisch N J 2010 *J. Appl. Phys.* **108** 093307.
[9] Smirnov A, Raitses Y, and Fisch, N J *Phys. Plasmas* **14**, 057106.
[10] Seo M, Lee J, Seon J, Lee H J and Choe, W 2013 *Phys. Plasmas* **20** 023507.
[11] Smirnov A, Raitses Y, and Fisch N J 2004 *Phys. Plasmas* **11** 4922.
[12] Matyash K, Schneider R, Kalentev O, Raitses Y and Fisch N J 2011 *The 32nd Intl. Elec. Prop. Conf.* IEPC-2011-070.
[13] Raitses Y, Smirnov A, and Fisch N J *Phys. Plasmas* **16** 057106.
[14] Ferreira J L, Martins A A, Miranda R A, Schelin A B, Alves L de S, Costa E G, Coelho E O, Serra A C B and Nathan F 2015 *J. Comp. Appl. Math* DOI 10.1007/s40314-015-0286-4.
[15] Pryor, R W 2015 *Multiphysics Modeling Using Comsol 5 and MATLAB* (Mercury Learning and Information).
[16] Choi Y, Mahalingam S, Likhanskii A. 2012 *IEEE Conf. Series* 978-1-4577-0557-1/12.
[17] Hockney R W and Eastwood J W 1988 *Computer Simulation Using Particles* (Bristol, IOP Publishing Ltd.).
[18] Taccogna F, Longo S, Capitelli M, Schneider R *Phys. Plasmas* **12** 053502.
[19] Hagelaar, G J M, Bareilles, J, Garrigues, L and Boeuf, J-P 2003 *J. Appl, Phys* **93** 67.
[20] Hagelaar, G J M 2007 *Plasma Sources Sci. Technol.* **16** S57.
[21] Hofer, R R, Katz, I, Mikellides, I G, Goebel, D M, Jameson, K K, Sullivan, R M and Johnson, L K. 2008. *AIAA/ASME/SAE/ASEE Joint Propulsion Conf. Exhib.* (Hartford), AIAA 2008-4924.
[22] Cho S, Komurasaki K, Arakawa Y 2013 *Phys. Plasmas* **20**, 063501.
[23] Boniface C, Garrigues L, Hagelaar G J M, Boeuf J P, Gawron, D and Mazouffre S 2006 *Appl. Phys. Lett.* **89**, 161503.
[24] Bultinck E, Mahieu S, Depla D, Bogaerts A 2010 *J. Phys D: Appl. Phys*, **43** 292001.

Orbital Trajectories in Deimos Vicinity Considering Perturbations of Gravitational Origin

L D Gonçalves[1], E M Rocco[1] and R V de Moraes[1, 2]

[1] National Institute of Space Research, INPE, São José dos Campos, Brazil
[2] Federal University of São Paulo, UNIFESP, São José dos Campos, Brazil

E-mail: lianadgon@gmail.com, evandro.rocco@inpe.br, rodolpho.vilhena@gmail.com

Abstract. The present work studies the possibility of maintaining an artificial satellite near to the Deimos surface, even when considering the intense Mars gravitational attraction. For this, a strategy is used in which the spacecraft orbits Mars, but with the same orbital period of Deimos and similar orbital elements. Using this strategy, the deviations in the keplerian elements that characterize the artificial satellite orbit, the magnitude of the main disturbing forces capable of altering the movement of the spacecraft and the variation of the distance between the spacecraft and Deimos are presented.

1. Introduction

Although smaller and more distant from Mars when compared to Phobos, Deimos was the first of the two moons of Mars discovered. Due to the proximity between Mars and Deimos, as well as the significant difference between their masses, the planet exerts a strong influence on the trajectory of an artificial satellite in the Deimos vicinity.

Thus, the present work aims to present a strategy to maintain an artificial satellite near to Deimos for a long period of time, even when the intense gravitational attraction of the Deimos is considered. For this study the Spacecraft Trajectory Simulator – STRS, [1] was used.

A study of the magnitude of the major disturbing forces capable of altering the orbital motion of an artificial satellite in the Deimos vicinity is made, as well as an analysis of the orbital elements behavior of the artificial satellite when disturbed by such forces.

For the attraction due to the Mars gravitational potential, the model presented by [2], which allows us to expand the spherical harmonics up to degree and order 80 is used. The non-uniform mass distribution of Deimos, its irregular shape, and consequently the non-central gravitational field, is modeled using the polyhedral method for the mass distribution, elaborate from data provided by NASA (National Aeronautics and Space Administration) [3].

2. Mars Gravitational Potential

The gravitational potential U of Mars is modeled in spherical harmonics from Equation (1), [4].

$$U = \frac{GM}{r}\left\{1 + \sum_{l=2}^{\infty}\left(\frac{a_e}{r}\right)^{l}\sum_{m=0}^{l}[\bar{C}_{lm}\cos m\lambda + \bar{S}_{lm}\sin m\lambda]\bar{P}_{lm}(\sin\phi)\right\} \tag{1}$$

where G is the gravitational constant, M is the mass of the planet, a_e is the equatorial radius, $\bar{P}_{lm}$ are the Legendre polynomials standardized for degree l and order m, r is the radial coordinate fixed in the body, ϕ is the latitude and λ is the longitude, and $\bar{C}_{lm}$ and $\bar{S}_{lm}$ are standardized coefficients for the expansion of spherical harmonics.

Different works using models that use the expansion of Equation (1), for high degree and order values for the expansion of the spherical harmonics for the Moon and for Mars, can be found in [5], [6], [7] and [8].

From the results obtained by the Mars Global Surveyor, the model GMM-2B was developed by [2]. This model is a representation of the spherical harmonics due to the planetary gravity based on the gravitational potential of the celestial body, given by Eq. (1). The model GMM-2B, used in this work, allows us to adopt as maximum precision of the model, the value 80 for degree and order of the spherical harmonics expansion.

3. Deimos Gravitational Potential

For the non-central gravitational potential generated by Deimos a polyhedral model developed by NASA (National Aeronautics and Space Administration [3]) was used, which represents the non-uniform mass distribution of the moon.

Based on the polyhedral model and on the methodology used by [9] and [10], was possible to define the mass concentrations corresponding to the mass of each polyhedron, being possible to calculate the gravitational potential generated by each concentration along the trajectory of the satellite disturbed by Deimos.

4. Strategy to Maintain the Spacecraft in the Deimos Vicinity

Due to the significant difference of mass between Deimos and Mars, and the fact that Deimos perform an orbit near to Mars, the planet's gravitational field exerts significant influence on the orbit of an artificial satellite around Deimos, making it impossible to maintain a satellite in a stable orbit for a long period of time. The physical and orbital parameters of Mars and Deimos are presented in Tables 1 and 2.

Table 1. Mars Parameters

Mass	0.64171×10^{24} kg
Equatorial radius	3396.2 km
Hill radius	22.9 km

Table 2. Deimos Parameters

Mass	1.8×10^{15} kg
Equatorial radius	6.2 ± 0.18 km
Semi major axis	23463.2 km
Eccentricity	0.00033
Inclination (with respect to the Mars equator)	0.93°
Orbital period	1.263 day

The influence sphere radius (Hill radius) presented in Table 1 was calculated from [11]. Due to the fact that Deimos orbits close to Mars, and considering its small mass, the value of the Hill radius to the moon is approximately 22.9 km. Thus, there is a significant influence of the Mars gravitational potential on an artificial satellite in the Deimos vicinity.

Thus, the strategy to maintain the artificial satellite close to Deimos adopted in this work, is to maintain the satellite in orbit of Mars, but with the same orbital period of Deimos and initial conditions similar to Deimos orbit around Mars. A study using this approach, for a satellite in the vicinity of Phobos, can be found in [12] and [13]. For this approach was used a reference system similar to that adopted in rendezvous and docking maneuvers. The coordinate axes H-bar, R-bar and V-bar shown in Figures 2 to 4 represent, respectively, the coordinates in the opposite direction of the angular momentum, in the direction of the center of the central body and in the direction of the velocity vector [14].

In Figure 1 the red orbit is the orbit of the artificial satellite around Mars and the green orbit is the orbit of Deimos. We can clearly see that the satellite orbits Mars, so it is possible to maintain it close to Deimos. This study was accomplished for a simulation time of 5 days, 30 days and 100 days.

From Figures 2 to 4 we can see that the satellite remains very close to Deimos, but over time it is moving away. This is because Deimos and the satellite are disturbed by Mars, but the satellite is also disturbed by Deimos. Due to this disturbance of Deimos, the satellite ceases to have the similar keplerian elements to the moon. Therefore, the disturbance due Mars, added to the disturbance due Deimos, causes the distance between the satellite and Deimos to increase over the time.

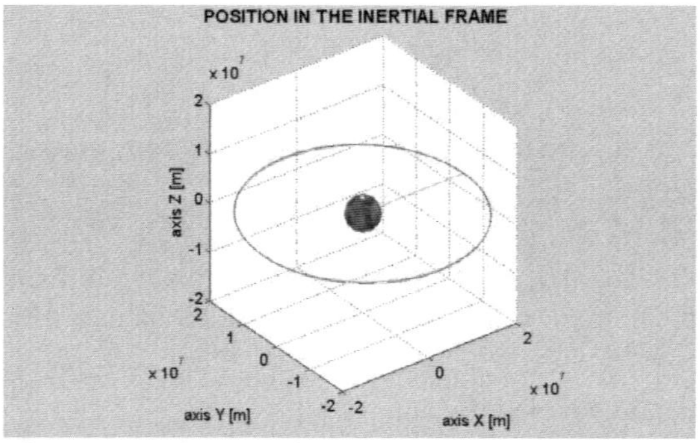

Figure 1. Deimos and satellite orbits around Mars.

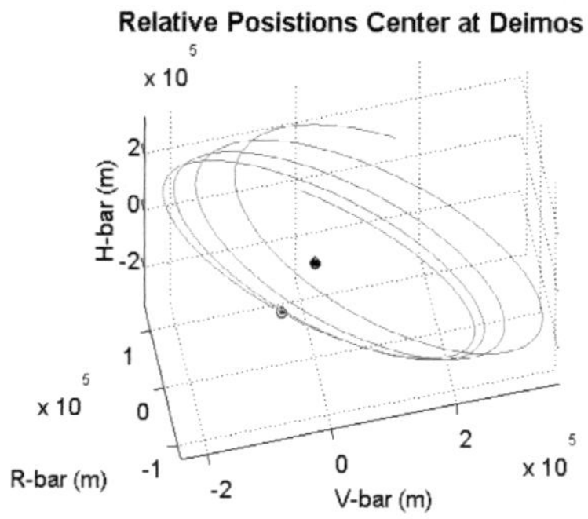

Figure 2. Satellite trajectory in the vicinity of Deimos (5 days).

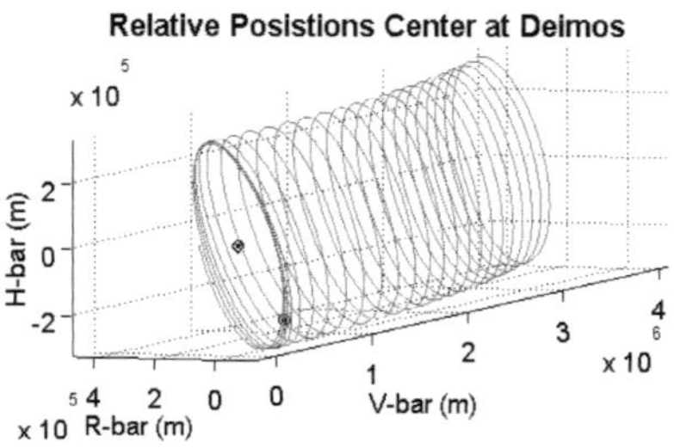

Figure 3. Satellite trajectory in the vicinity of Deimos (30 days).

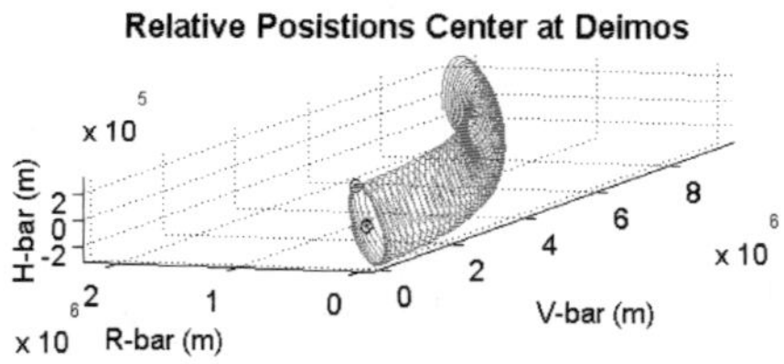

Figure 4. Satellite trajectory in the vicinity of Deimos (100 days).

In Figures 5 to 7 the keplerian elements that characterize the orbit of the artificial satellite are presented, only for the case of simulation time of 100 days.

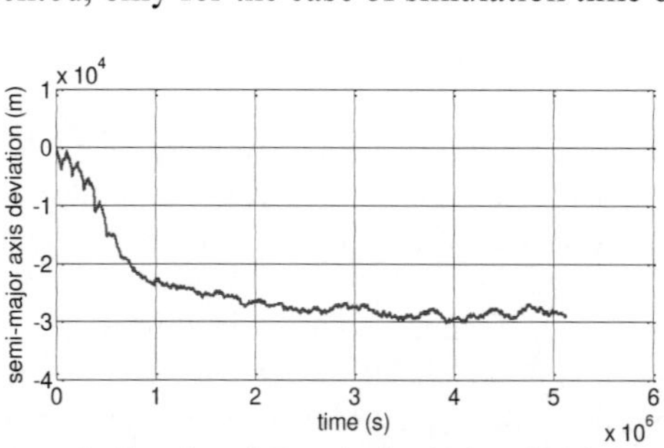

Figure 5. Semi-major axis deviation (100 days).

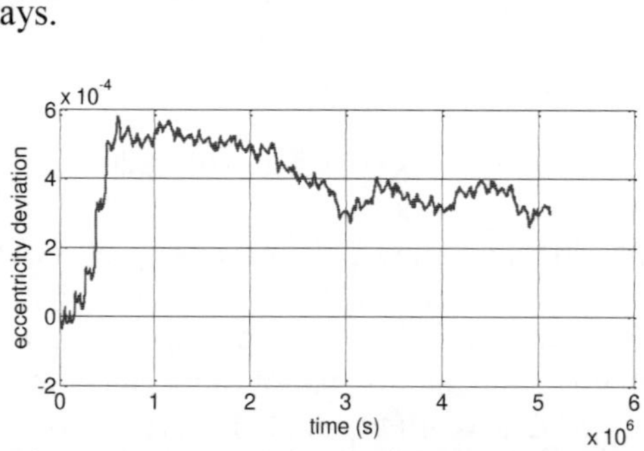

Figure 6. Eccentricity deviation (100 days).

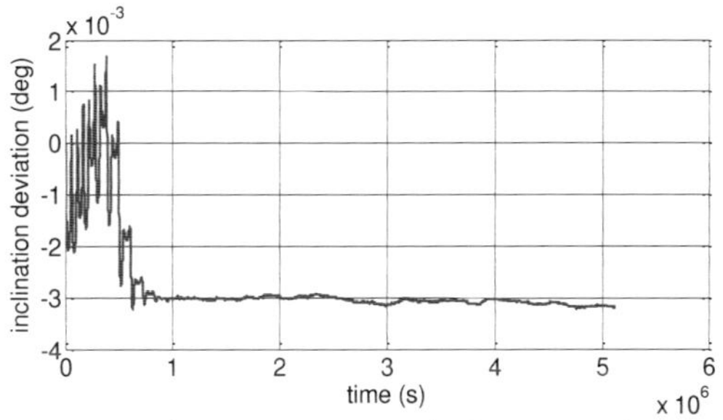

Figure 7. Inclination deviation (100 days).

Figures 5 and 7 show that the deviation in the semi-major axis and the inclination were more pronounced at the beginning of the simulation, when the satellite was closer to Deimos. Due to the effect of the disturbances, as the satellite moves away from Deimos, the deviation in the eccentricity increases. It is worth noting that the deviations presented are to the satellite orbit around Mars, following the proposed approach.

Figures 8 to 15 show the perturbation velocity increment due to Deimos, Mars, the total perturbation and the distance between the spacecraft and Deimos for the simulation time of 5 and 30 days, and Figures 16 to 20 show the increment of velocity Disturbing due to Deimos, Mars, Sun, the total disturbance on the spacecraft and the distance between the satellite and Deimos for the simulation time of 100 days.

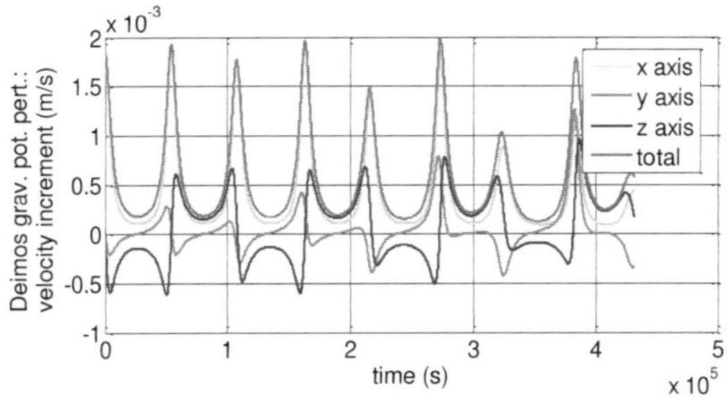

Figure 8. Distrubance due Deimos (5 days).

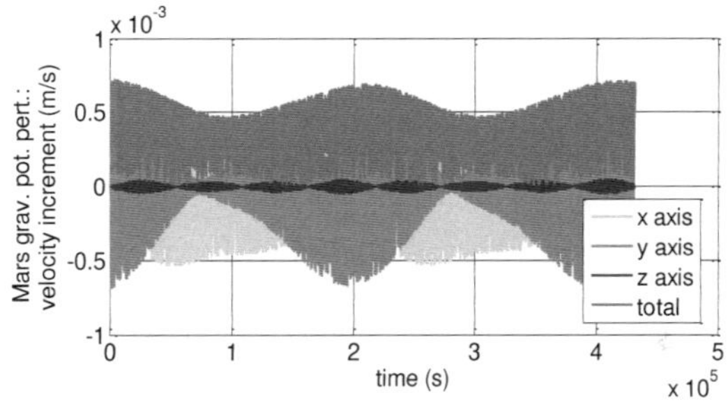

Figure 9. Distrubance due Mars (5 days).

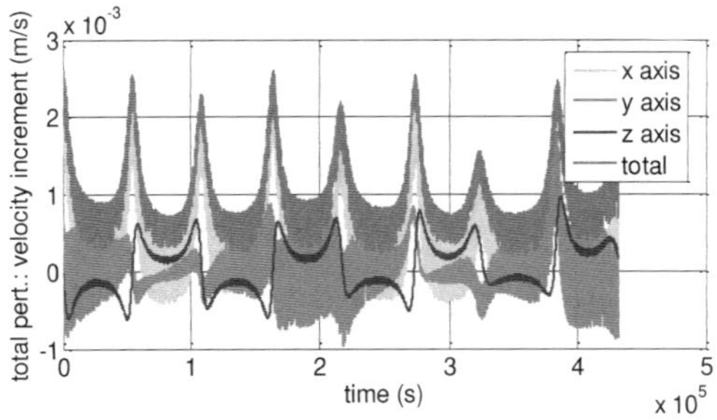

Figure 10. Total distrubance (5 days).

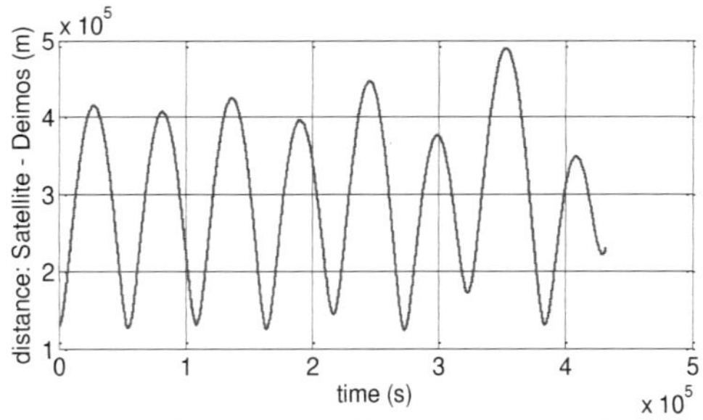

Figure 11. Distance satellite-Deimos (5 days).

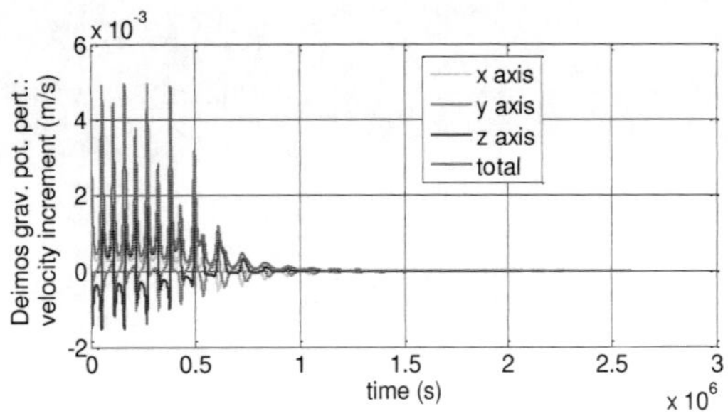

Figure 12. Distrubance due Deimos (30 days).

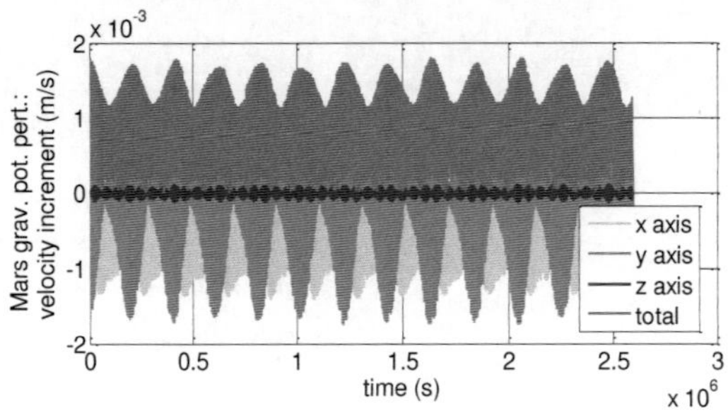

Figure 13. Distrubance due Mars (30 days).

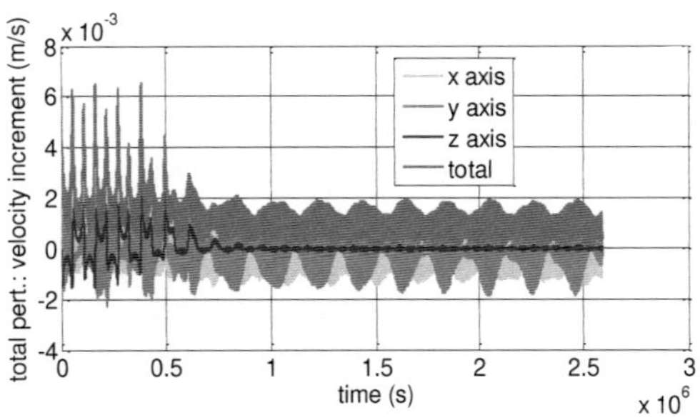

Figure 14. Total distrubance (30 days).

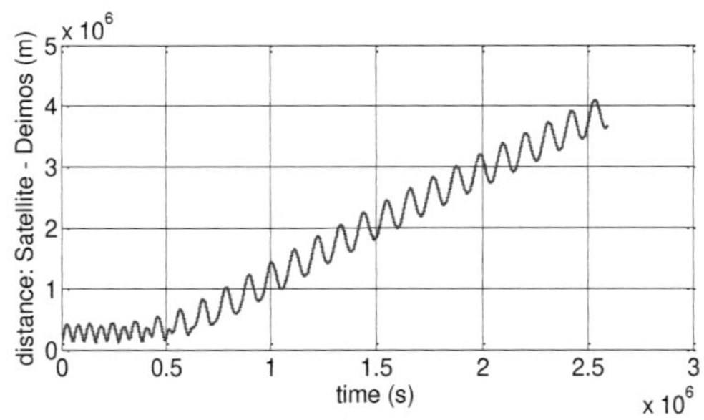

Figure 15. Distance satellite-Deimos (30 days).

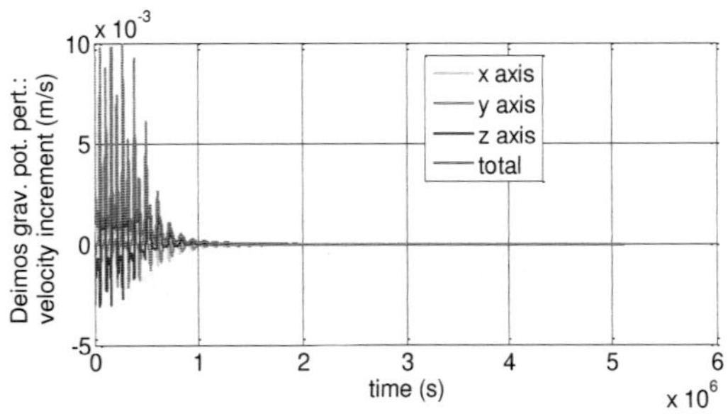

Figure 16. Distrubance due to Deimos (100 days).

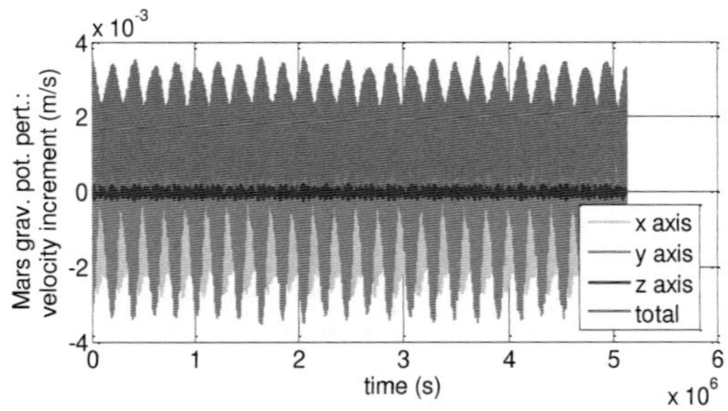

Figure 17. Distrubance due to Mars (100 days).

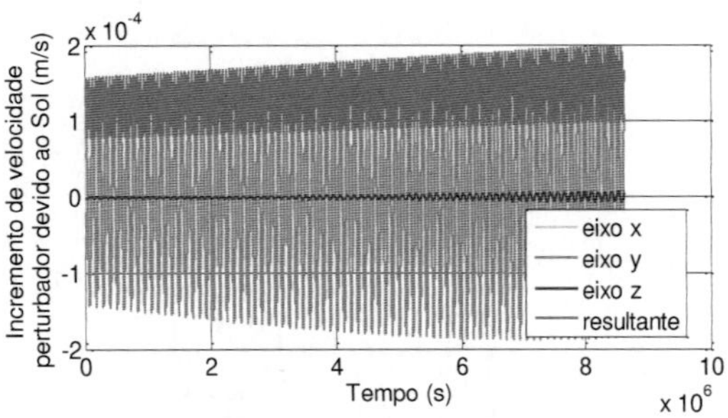

Figure 18. Distrubance due to Sun (100 days).

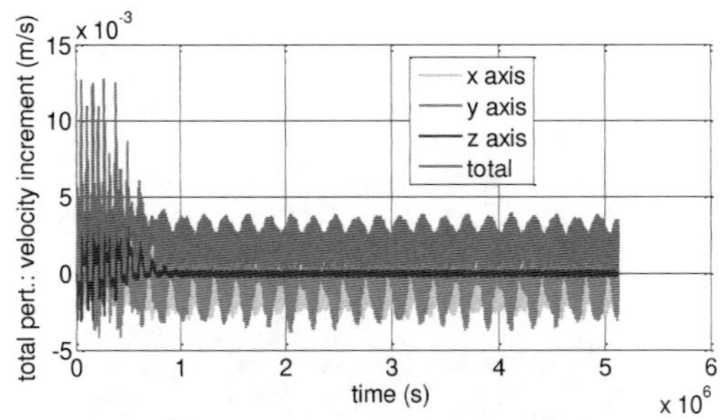

Figure 19. Total distrubance (100 days).

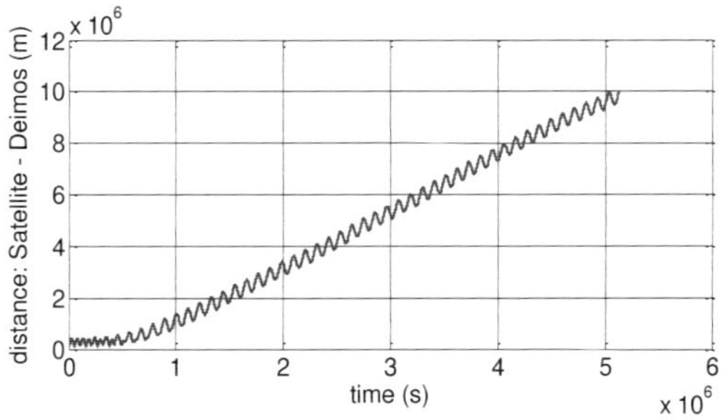

Figure 20. Distance satellite-Deimos (100 days).

It can be noticed difference in magnitude of the perturbation due to the gravitational potential of the studied bodies. Even with significantly higher mass, in the case of the initial conditions studied, Mars exerts less influence than Deimos. However, such a difference is not enough to maintain the spacecraft orbiting Deimos indefinitely, and the sum of the disturbances causes the orbital elements to evolve over time and the spacecraft distances itself from the moon, as seen in Figures 15 and 20. The inconstancy in the gravitational potential of Mars (Figures 9, 13 and 17) is also notable. This is because is considered high degree and order values in the expansion of the spherical harmonics of the planet.

When Figures 16 and 20 (and the respectives figures for the case of 5 and 30 days) are compared, is possible to observe that there are peaks of greater perturbation due to the gravitational potential of Deimos. This is due to the fact that there are moments of closer approximation between the spacecraft and Deimos, as seen, for example, in Figures 16 and 20.

5. Conclusions

The perturbations studied have close magnitudes, which makes difficult to maintain a spacecraft orbiting Deimos. However, with the adopted approach is possible to realize trajectories near to Deimos surface, even when considered the Mars gravitational attraction. Was verified the variation in the orbital elements, mainly in the semi-major axis and eccentricity due essentially to the Deimos non-central gravitational potential and the Mars gravitational attraction. The approach of the spacecraft to Deimos, as well as to maintain the spacecraft in this region could be of great importance for observation, conducting experiments or even landing.

References

[1] Rocco E M 2015 Gravitational disturbances generated by the Sun, Phobos and Deimos in orbital maneuvers around Mars with automatic correction of the semi-major axis. *Journal of Physics Conference Series*, v. 641. doi:10.1088/1742-6596/641/1/012027

[2] Lemoine F G; Smith D E; Rowlands D D; Zuber M T; Neumann G A; Chinn D S; Pavlis D E 2001 An improved solution of the gravity field of Mars (GMM-2B) from Mars Global Surveyor. *Journal of Geophysical Research*, Vol 106.

[3] Gaskell R W 2011 Gaskell Phobos Shape Model V1.0. VO1-SA-VISA/VISB-5-PHOBOSSHAPE-V1.0. NASA Planetary Data System.

[4] Kaula W M 1966 *Theory of Satellite Geodesy*, Blaisdell, Waltham, Mass.

[5] Gonçalves L D; Rocco E M; de Moraes R V; Prado A F B A 2016 Effects of the individual terms of the lunar potential in the motion of satellites around the Moon. *International Journal of Theoretical and Applied Mechanics*, v. 1, p. 20-29

[6] Gonçalves L D; Rocco E M; de Moraes R V 2015 Analysis of the influence of orbital disturbances applied to an artificial lunar satellite. *Journal of Physics. Conference Series* (Print), v. 641, p. 012028, 2015

[7] Gonçalves L D; Rocco E M; de Moraes R V; KUGA H K 2015 Kalman filter application to mitigate the errors in the trajectory simulations due to the lunar gravitational model uncertainty. *Journal of Physics. Conference Series* (Print), v. 641, p. 012029

[8] Gonçalves L D; Rocco E M; de Moraes R V 2013 Orbital Disturbance Analysis due to the Lunar Gravitational Potential and Deviation Minimization through the Trajectory Control in Closed Loop. *Journal of Physics Conference Series* (Print), v. 465, p. 012013

[9] Venditti F C; Rocco E M; Prado A F B A 2013 Estudo de Perturbações em Órbitas ao Redor do Asteroide 216 Kleopatra Utilizando Modelo de Poliedros 2013. DINCON - Conferência Brasileira de Dinâmica, Controle e Aplicações, Fortaleza. Anais do DINCON 2013 - Conferência Brasileira de Dinâmica, Controle e Aplicações. Fortaleza: SBMAC,. v. 1. p. 1-6.

[10] Rocco E M 2014 Trajetórias de aproximação para pouso no asteroide 216 Kleopatra. In: XVII Colóquio Brasileiro de Dinâmica Orbital - CBDO, Águas de Lindóia - SP - Brasil. Anais do XVII Colóquio Brasileiro de Dinâmica Orbital

[11] Bonneau F; Bernard J 1995 Interplanetary trajectories. In: Spaceflight Dynamics. Cépaduès Éditions

[12] Gonçalves L D; Rocco E M; de Moraes R V Trajetórias na vizinhança de Fobos visando sobrevoo a aproximação 2016. In: Colóquio Brasileiro de Dinâmica Orbital, 2016, Águas de Lindoia. Resumos do XVIII CBDO

[13] Rocco E M; Gonçalves L D; de Moraes R V 2017 Orbital Maneuvers in The Non-Central Gravitational Fiedl of Phobos Perturbed by the Gravitational Attraction due to Mars, Sun and Deimos. In: 27th Space Flight Mechanics Meeting, San Antonio. Proceedings of the 27th Space Flight Mechanics Meeting

[14] Fehse W 2003 *Automated Rendezvous and Docking of Spacecraft*

IOP Conf. Series: Journal of Physics: Conf. Series **911** (2017) 012023 doi:10.1088/1742-6596/911/1/012023

Equilibrium points in the asteroid 2001SN$_{263}$

Santos, L. B. T.[1], **Prado, A. F. B. A.**[1], **Sanchez, D. M.**[1]
[1]National Institute from Space Research, INPE, São José dos Campos, Brazil
E-mail: leonardo.btorres@inpe.br

Abstract. In the present work we develop a study based on the Restricted Full Three-Body Problem (RFTBP), where we take into account the size and shape of the bodies studied. The equations of motion and their solutions are discussed for a circular coplanar orbit in order to find the equilibrium points. All the analysis made are based on the physical and orbital characteristics of the most massive bodies of the triple asteroid system $2001SN_{263}$. A total of nine equilibrium points were found for this system. Five of these points are collinear and the other four are coplanar. Finally, an analysis of the stability conditions of these equilibrium points are investigated, and it was concluded that only the equilibrium points L_4 and L_5 are linearly stable.

1. Introduction

In the last years there has been a great interest in the exploration of small bodies of the solar system. Mathematical modeling of the gravitational field near the surface of an asteroid is a very challenging task, as it is necessary to know some physical properties of the asteroid, such as size, shape, density, etc, which are often only discovered after the spacecraft approaches the body [1]. To make an analysis of the orbital dynamics of a spacecraft around an irregular body, it is necessary to obtain a mathematical model to represent the gravitational field of the body that one wishes to orbit. Due to the fact that many asteroids have an irregular shape, the orbits of spacecraft near these bodies do not assume a traditional Keplerian orbit [2]. Due to the rotation effect of asteroids along with their irregular shapes, the gravitational fields acting on spacecraft, which are very close to these bodies, are very complex [3]. Then, when planning a space mission, it is necessary to make mathematical modeling flexible enough to cover several parameters of the object being studied [1]. In the present work an analysis was made considering the primary bodies as spheroids. The asteroids adopted to carry out this study are the asteroid *Alpha* and *Beta* of the triple asteroid system $2001SN_{263}$. This asteroid system was chosen to be analyzed due to a Brazilian mission known as Missao Aster [4, 5, 6, 7].

2. Equation of motion

Let be two primary bodies (M_1 and M_2) with finite mass (m_1 and m_2) and of spheroid format. Let us consider here that $m_1 > m_2$. Suppose that these bodies are rotating in a circular motion about the center of mass of the system with angular velocity ω. Now consider a third body ($P(x, y, z)$), and that the motion of a negligible mass is governed by the gravity forces of the bodies M_1 and M_2. The infinitesimal mass body does not affect the dynamics of the primaries, as shown in Figure 1. The unit of distance is normalized to be the distance from the body M_1 to the body M_2.

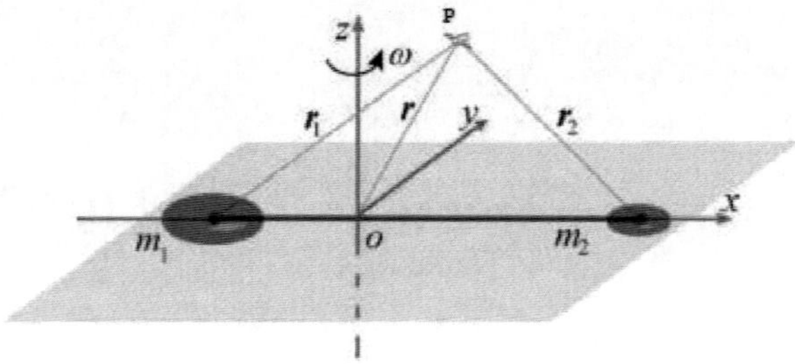

Figure 1: Representation of the position of the bodies m_1, m_2 and P with respect to the center of mass (O) of system in a synodic frame. Source: [3].

The unit of time is defined such that the period of translation of the primary bodies, is equal to 2π. Thus, we have $G(m_1 + m_2) = 1$. The spatial position of the negligible mass is $P(x, y, z)$ and the coordinates of the bodies with mass m_1 e m_2 are $(-x_1, 0, 0)$ and $(x_2, 0, 0)$, respectively, with respect to the rotating system. The mass ratio is given by $\mu = \frac{m_2}{m_1+m_2}$. The primary bodies are on the axis x, whose coordinates are given by $x_1 = -\mu$ and $x_2 = 1 - \mu$. The equations of motion of the negligible mass body, in the xy plane, when viewed from a rotating frame of reference, is given by the equations (1-6).

$$\ddot{x} - 2\dot{y} = U_x \tag{1}$$

$$\ddot{y} + 2\dot{x} = U_y \tag{2}$$

where

$$U = -\frac{\omega^2(x^2 + y^2)}{2} - k\omega^2 \left(\frac{(1-\mu)w_1}{r_1} + \frac{\mu w_2}{r_2} \right) \tag{3}$$

where

$$w_i = 1 + \frac{A_i}{2r_i^2}, i = 1, 2 \tag{4}$$

,

$$r_1 = [(x - x_1), y, 0]^T \tag{5}$$

$$r_2 = [(x - x_2), y, 0]^T \tag{6}$$

and U_x and U_y are the partial derivatives of U with respect to x and y, respectively.

The dimensionless parameter k that appears in Equation 3 is of extreme importance, and represents the ratio between gravitational force and centrifugal force, whose definition is given by:

$$k = \frac{GM/d^2}{d\omega^2} \tag{7}$$

where G is the universal gravitational constant, M is the sum of the masses of the primary bodies, and d is the distance between the primary bodies. The angular velocity in the dimensionless system that appears in Equation 3 is given by:

$$\omega = \sqrt{1 + \frac{3(A_1 + A_2)}{2}} \tag{8}$$

where A_i $(i = 1, 2)$ are the oblateness coefficient of the primary bodies, whose definition is given by:

$$A_i = \frac{(\rho_i^e)^2 - (\rho_i^p)^2}{5d^2} \tag{9}$$

IOP Conf. Series: Journal of Physics: Conf. Series **911** (2017) 012023 doi:10.1088/1742-6596/911/1/012023

where the parameter ρ is the radius of the primary spheroids. The subscript e represents the equatorial radius of the primary and the subscript p indicates the polar radius of the primary [3].

3. Results

In order to calculate the equilibrium points of a binary system of asteroids it is necessary to know the characteristics of this system. Recent data from the asteroid system $2001SN_{263}$ can be found in [8]. In this study, we will consider asteroids with spheroid shapes. The equatorial radius of the asteroid *Alpha* is estimated to be 1.45 km and the radius of the Polar axis is 1.35 km. If we take as reference the asteroid *Alpha* (because it is the most massive and largest body), we have that the radius of the polar axis of the asteroid *Beta* is 0.5 km and the radius of the equatorial axis of the asteroid *Beta* is 0.32 km. A geometric representation of the shape of the alpha and beta (not in scale) asteroids can be seen in Figure 2.

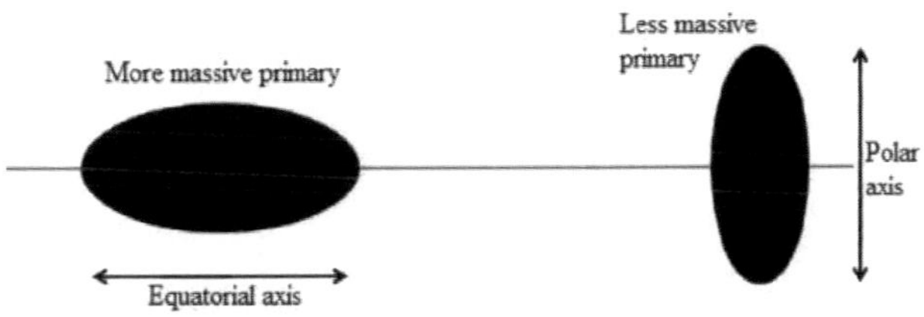

Figure 2: Spheroid format of a binary system of asteroids.

3.1. Equilibrium Points

To find the equilibrium points, it is necessary to make the right side of the Equations 1 and 2 equal zero. It is possible to note, by analyzing the Equation 1 and 2, that, if a particle is placed at an equilibrium point with zero initial velocity, it will remain at this point indefinitely.

Let the right side of Equations 1 and 2 to be equal zero, that is, $\frac{\partial U}{\partial x} = 0$ and $\frac{\partial U}{\partial y} = 0$, and replacing the numerical values of the analyzed system, it was possible to find nine equilibrium points in the orbital plane of the primaries. Therefore, in addition to the five classical equilibrium points, four new equilibrium points appeared. The Figure 3(a) shows the positions of the nine equilibrium points of the binary system of asteroids studied and the location of the center of mass of the primary bodies. Red represents the classic lagrangean points, known as L_1, L_2, L_3, L_4 and L_5. In blue are the positions of the center of mass of the primary bodies. The blue dot on the left represents the most massive primary (*alpha*) and the blue dot on the right represents the least massive primary (*Beta*). In green, very close to *Beta*, are the new equilibrium points called here L_6, L_7, L_8 and L_9.

Figure 3(b) is an approximation of the less massive body, for better visualization of the points L_6, L_7, L_8, and L_9. We see that these new equilibrium points are very close to the center of mass (blue color) of the body *Beta*.

When we consider the dimension of the less massive body, based on the dimensions of the asteroid *Beta*, we see that the new points of equilibrium are inside the less massive body, as shown in Figure 4(a). In this figure we can see an oval representing the asteroid *Beta* (violet color), the center of mass of this asteroid (blue color) and the new equilibrium points (green color).

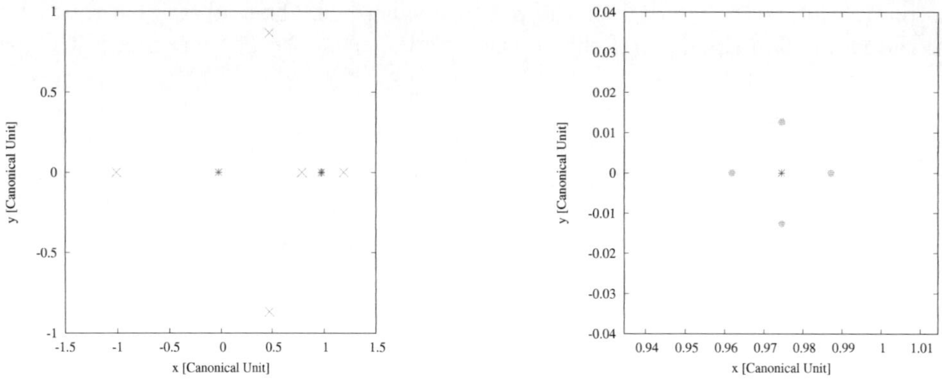

(a) Image of the nine equilibrium points showing the classical points (red color), non-classical points (green color) and the positions of the center of mass of the primary bodies (blue color).

(b) New equilibrium points, called L_6, L_7, L_8 and L_9 (green color) and the center of mass of the less massive asteroid (blue color).

Figure 3: Equilibrium points.

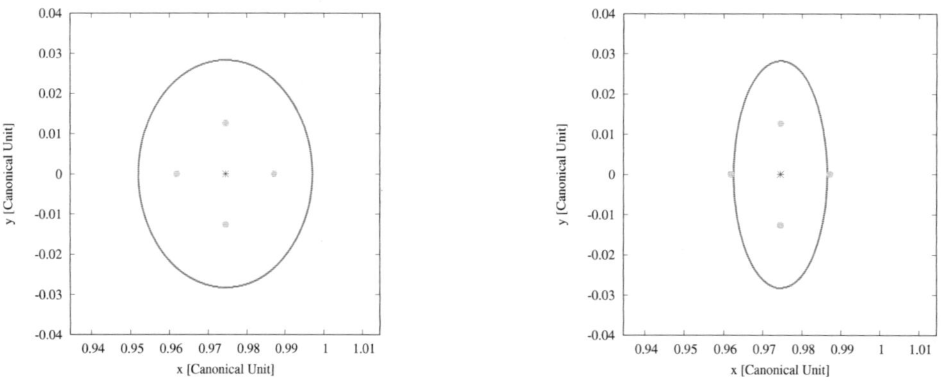

(a) New equilibrium points, called L_6, L_7, L_8 and L_9 (green color), the center of mass of the less massive asteroid (blue color) and an oval (violet color) representing the dimension of the less massive asteroid.

(b) New equilibrium points, called L_6, L_7, L_8 and L_9 (green color), the center of mass of the less massive asteroid (blue color) and an oval (violet color) representing the dimension of the less massive asteroid (taking into account the margin of error).

Figure 4: Equilibrium points and size of asteroid Beta.

However, when we take into account the error margins of the dimension of this asteroid, some of these new points of equilibrium are practically on the surface of this asteroid, as shown in Figure 4(b).

This is quite interesting, because there is a point of equilibrium practically on the surface of an asteroid. If a spacecraft has an interest in landing on a binary system of asteroids with specifications similar to those discussed here, it will be possible to find a position on the asteroid where the spacecraft can reach the surface with zero velocity and thus be able to land.

To know where these stationary points are located is important for, as an example, maintaining a spacecraft, since these points indicate the locations where the spacecraft suffers the least disturbance and thus the minimum possible fuel is spent for station-keeping.

The distances of the lagrangean points calculated here are distances relative to the center of mass of the system. Table 1, of the studied system, is normalized, considering the distance between the two primaries as being unitary.

Table 1: Position of the equilibrium points when taking into account the oblateness of the bodies.

	L_1	L_2	L_3	L_4	L_5	L_6	L_7	L_8	L_9
x	0.78438	1.19238	-1.01072	0.47466	0.47466	0.96185	0.98716	0.97452	0.97452
y	0	0	0	0.86605	-0.86605	0	0	0.01265	0.01265

3.2. Stability

Let the coordinates of some equilibrium points be given by $(x_0,\ y_0,\ 0)$. We are considering $z = 0$ because we are analyzing the equilibrium points in the plane. Now let us suppose that α and β are small displacements of a particle of infinitesimal mass with respect to the point of equilibrium, in the directions x and y, respectively. These displacements are functions of time, and therefore there will be acceleration components $(\ddot{\alpha},\ \ddot{\beta},\ 0)$ and velocity $\dot{\alpha},\ \dot{\beta},\ 0$.

In this way, we can write the equation of the motion of a particle of infinitesimal mass as follows:

$$\ddot{\alpha} - 2\dot{\beta} = \alpha(U_{xx})_0 + \beta(U_{xy})_0, \tag{10}$$

$$\ddot{\beta} + 2\dot{\alpha} = \alpha(U_{yx})_0 + \beta(U_{yy})_0. \tag{11}$$

Equations 10 and 11 are the linearized equations of motion of a particle of infinitesimal mass in the vicinity of a point of equilibrium when it is slightly displaced [9, 10, 11]. To kwnow if the equilibrium point is linearly stable or not, simply replace the positions of theses points in Equations 10 and 11 and solve the system.

If all eigenvalues of the equations give only purely imaginary values, this implies that the equilibrium point analyzed will be stable up to first order (or linearly stable). On the other hand, if any of the eigenvalues is real or imaginary, it is an unstable equilibrium point [9, 10, 11].

Using Equations 10 and 11, adopting the physical and orbital properties of *Alpha* and *Beta*, and knowing the values of the equilibrium points, it was possible to analyze the stability of all Equilibrium points found, based on the eigenvalues, as shown in Table 2.

Table 2: Eigenvalues of the characteristic equation for the RFTBP.

	L_1	L_2	L_3	L_4	L_5	L_6	L_7	L_8	L_9
λ_1	-3.045	-2.055	0.258	$0.460i$	$0.460i$	111.965	112.199	0.166	112.25
λ_2	3.045	2.055	-0.258	$-0.460i$	$-0.460i$	-111.965	-112.199	-0.166	-112.25
λ_3	$2.412i$	$1.807i$	$1.021i$	$0.886i$	$0.886i$	$274.374i$	$274.836i$	$1.009i$	$274.918i$
λ_4	$-2.412i$	$-1.807i$	$-1.021i$	$-0.886i$	$-0.886i$	$-274.374i$	$-274.836i$	$-1.009i$	$-274.918i$

We can note that only the equilibrium points L_4 and L_5 are linearly stable. The other points are all unstable.

4. Conclusion

In the Equations of motion the RFTBP, when we considered the oblateness of the primary bodies, new equilibrium points were found. In addition to the five classical equilibrium points found in the restrited circular/elliptic three-body problem, four new coplanar equilibrium points appeared (here called L_6, L_7 , L_8 and L_9). This is due to oblateness of the bodies. When we consider the asteroid data of *Alpha* and *Beta*, we realize that *Alpha* is an oblate body (it means, more elongated in the equatorial axis) and the body *Beta* has a prolate oblateness (it is more elongated in the polar axis). This prolate form gives rise to new equilibrium points around the body with a greater elongation in the polar axis.

An analysis of the stability of all these points found (L_1 to L_9) was also made and the numerical evidence shows that only the equilibrium points L_4 and L_5 are linearly stable, while all others are unstable. The equilibrium points of the asteroid system are considered to be places that receive minimal disturbances, so they are good places to maintain a spacecraft. It can be used as parking orbits. It can be interesting to keep a spacecraft near the surface of an asteroid, thus being a suitable place to start landing. So, the points found here are very interesting for this purpose.

5. acknowledgments

The authors wish to express their appreciation for the support provided by grants# 406841/2016-0 and 301338/2016-7 from the National Council for Scientific and Technological Development (CNPq). We also thank the grants# 2016/14665-2, 2016/18418-0, 2011/08171-3, 2014/22293-2, 2014/2295-5 from São Paulo Research Foundation (FAPESP) and the financial support from the National Council for the Improvement of Higher Education(CAPES).

References

[1] SCHEERES, D. J.; WILLIAMS, B. G.; MILLER, J. K. Evaluation of the dynamic environment of an asteroid: Applications to 433 eros. **Journal of Guidance, Control and Dynamics**, v. 23, n. 3, p. 466475, 2000.

[2] MASAGO, B. Y. P. L. et al. Developing the -precessing inclined bi-elliptical four-body problem with radiation pressure- to search for orbits in the triple asteroid 2001sn263. **Advances in Space Research.**, v. 57, p. 962982, 2016.

[3] ZENG, X. Y.; GONG, S.; LI, J.; ALFRIEND, K. Equilibrium points of elongated celestial bodies as the perturbed rotating mass dipole. **Journal of Guidance, Control and Dynamics**, v. 39, n. 6, p. 1223, 2016.

[4] SUKHANOV, A. A., Velho, H.F.C., Macua, E.E., Winter, O.C.: **Cosmic research** 48(5), 455 (2010).

[5] PRADO, A. F. B. A. Mapping Orbits Around the Asteroid 2001SN263. **Advances in Space Research.**, v 53, p. 877, 2014.

[6] ARAJO, R. A. N., WINTER, O. C., PRADO, A. F. B. A., and Prado, A. F. B. A., VIEIRA, M. R. Sphere of influence and gravitational capture radius: a dynamical approach. **MNRAS**, v 391, p. 675, 2008.

[7] ARAJO, R. A. N., WINTER, O. C., PRADO, A. F. B. A., and Prado, A. F. B. A. Stable retrograde orbits around the triple system 2001 SN263. **MNRAS**, v 153, p. 1143, 2015.

[8] TRACY, M. B. et al. Physical modeling of triple near-earth asteroid (153591) 2001SN263 from radar and optical light curve observations. **Icarus**, v. 248, p. 499515, 2015.

[9] SZEBEHELY, V. **Theory of Orbits**. New York and London: Academic press., 1967.

[10] MCCUSKEY, S. W. **Introduction to Celestial Mechanics**. 1. ed. USA: Addison-Wesley Publishing Company, 1963.

[11] SANTOS, L. B. T., PRADO, A. F. B. A., SANCHEZ, D. M. Equilibrium points in the restricted synchronous three-body problem using a mass dipole model **Astrophysics and Space Science.** 362(61), 60 (2017).

IOP Conf. Series: Journal of Physics: Conf. Series **911** (2017) 012024 doi:10.1088/1742-6596/911/1/012024

Magnetic Field Design for a Strongly Improved PHALL Thruster

Alexandre A Martins[1], Rodrigo Miranda[1,2], José Leonardo Ferreira[1]

1- Plasma Physics Laboratory, University of Brasilia, 70910-900 Brasilia-DF, Brazil.
2- Faculty of Technology of UnB at Gama-DF, Brazil.

Email: aam@ist.utl.pt

Abstract. In this article, we are going to go through some steps that we took in the refining of engineering work related to the development of a permanent magnet Hall thruster. The use of permanent magnets in these thrusters is mainly related to the decrease of used power for propulsion, especially important for low power thrusters as for micro-satellites. The advantage of our chosen configuration is that the magnetic field can be used either perpendicular or parallel to the thruster channel walls, whereas in the last case the generated erosion forces are strongly reduced by at least three orders of magnitude. We are going to show how each magnetic field configuration affects the generated plasma and consequently the generated propulsion force and efficiency.

1. Introduction

Since 2004, the Plasma Physics Laboratory (LFP) of University of Brasilia - UnB is developing high efficiency Hall Effect Plasma Thrusters (PHALL) using permanent magnets [1-5]. These types of thrusters are now currently being used for commercial and scientific space missions for satellite attitude and orbit control and for long duration space missions in the solar system. In order to simulate the vacuum conditions of the space environment, high vacuum systems are used at LFP, in order to test the thrusters in conditions similar to those that are encountered in space, where they will work.

These thrusters have an anode in the form of a metallic ring that is inside a ceramic annular channel. A vertical ion accelerating electric field is generated between this anode and a virtual cathode made of an electron Hall current generated upwards to the anode in a region where the electric and magnetic fields are mutually perpendicular. The electrons are strongly magnetized by this magnetic field and follow a circular unidirectional path determined by the ExB fields. Ions are accelerated towards this virtual cathode and are ejected from the thruster at high velocity generating a reaction thrusting force.

Generally, the used magnetic fields are perpendicular to the vertical annular ceramic walls of the thruster, generating also a strong erosion process of the protective thruster material in a relatively short time frame severely limiting the lifetime of the thruster. This occurs because the electrons are also strongly directed along magnetic field lines towards the walls, leading the ions to follow their path by electrostatic attraction. However, there is a new and much better magnetic field configuration known as "magnetic shielding" where the magnetic fields are parallel to the vertical walls and significantly increase the lifetime of our thrusters for long duration space missions.

We will show for the first time how we have used ferromagnetic materials in order to control the magnetic field intensity of permanent magnets, which also allow for a decrease of the magnetic field to the desired intensities while helping at the same time to strongly making these fields much more uniform. In this way, we can engineer the shape of the thruster's magnetic fields in order to allow the correct and controlled development of Hall currents along uniform lines of perpendicular ExB spaces. When the magnetic shielding configuration is used we are able to generate a circular Hall current just outside the exit channel of the annular thruster therefore avoiding most of the contact and interaction between the generated plasma and the ceramic walls increasing dramatically the thruster's working lifetime and operational usefulness.

2. Magnetic Field Simulation Results for B Field Perpendicular to Wall and their Effects on Plasma Current

Initially, the intensity of the magnetic field **B** at the center of the internal channel of the PHALL II-A, where the plasma discharge occurs, was around 700-900 Gauss (Figure 1) in the whole vertical length of the 35.5 mm magnet height. This configuration caused a strong magnetization of both ions and electrons (which inhibited the generation of propulsive forces) generating a plasma torus (Figure 2). This magnet configuration had a strong gradient and non-uniformity of the **B** field along the internal channel of PHALL due to the aluminum support (non-ferromagnetic) structure that was initially used.

This initial result shows the need to use magnetic fields of specific magnitudes that can magnetize the electrons but not the ions. Magnetic fields in the order of 100 Gauss satisfy the conditions for length scales in the order of 10 mm. In practice, conventional Hall thrusters with powers in the order of kilowatts use magnetic fields with average peak magnitude around 100-300 Gauss at the center of the annular channel, with a depth and width around 25 mm.

For this propulsion system to work we have to use conditions where the Larmor radius of the electrons (r_e) is substantially smaller than the dimensions of the annular channel (L), which itself has to be substantially smaller than the Larmor radius of the ions (r_i) in such a way that the ions are not magnetized and can be ejected from the thruster in order to generate a useful propulsive force [6]. These relations are explicit in Equation 1 (where v_e and v_i is the electron and ion velocity respectively, e is the electron elemental charge and B is the magnetic field magnitude) and Table 1.

$$r_e = \frac{m_e v_e}{eB} < L < r_i = \frac{m_i v_i}{eB}. \quad (1)$$

Table 1. Magnetic field intensity and Larmor radius for ions and electrons depending on applied magnetic field.

Magnetic Field Strength (Gauss)	**r_i (mm) (1000 K)**	**r_e (mm) (1 eV)**	**r_e (mm) (10 eV)**
1	5500	40	120
100	55	0.4	1.2
1000	5.5	0.04	0.12

This ion magnetization problem was solved by replacing the initial long magnets (10x10x35.5 mm) with other shorter and smaller magnets (10x10x10 mm, Figure 3) and by replacing the initial aluminum support for the magnets with a different support with a material of ferromagnetic nature (Figures 3 and 4). By simply decreasing the magnet length and using these shorter magnets with no ferromagnetic support we can generate immediately lower fields (100-500 Gauss) than before (700-900 Gauss), but 500 Gauss (Figure 3) is still too high for our purposes (Table 1). By adding the ferromagnetic support into the mix this solution allows for the necessary decrease of magnetic field intensity inside the channel to levels where only the electrons are magnetized (100-300 Gauss) as desired, and on the other side it also allows for a higher uniformization of magnetic field lines inside the channel (Figures 4 and 5).

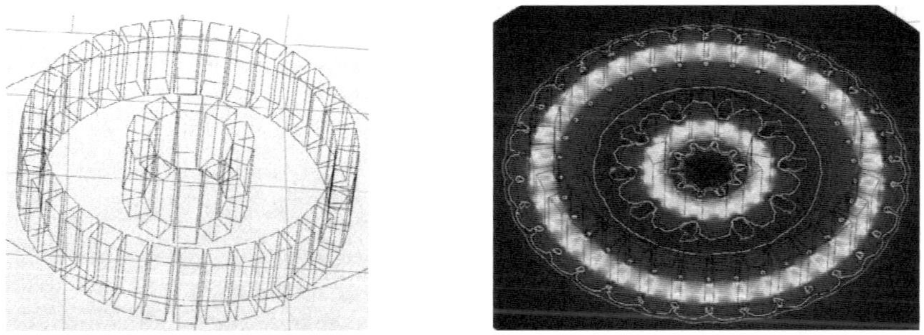

(a) (b)

Figure 1. Initial magnetic field configuration of PHALL II-A with no ferromagnetic material (magnets with length of 35.5 mm): a) 3D structure of magnets, b) Magnetic field line map for lines with 700 (green), 800 (yellow) and 900 (red) Gauss.

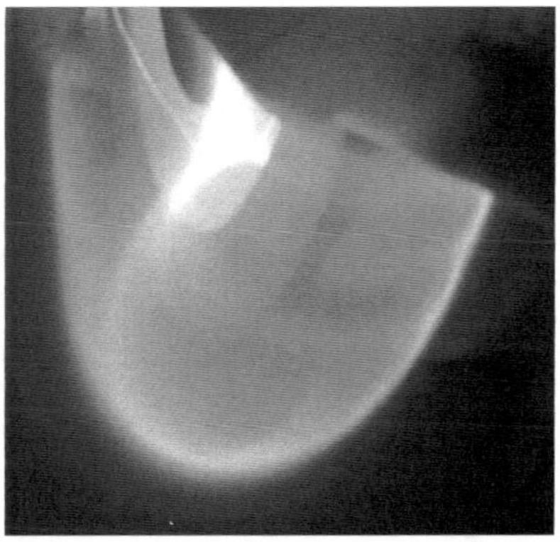

Figure 2. Plasma torus due to strong magnetization of both ions and electrons.

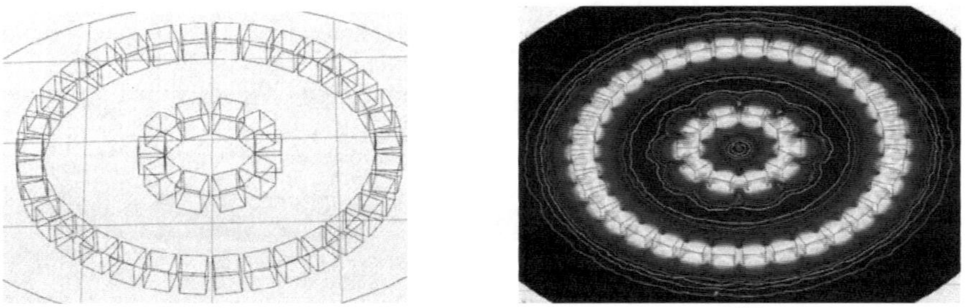

(a) (b)

Figure 3. Magnetic field configuration of PHALL II-B with no ferromagnetic material reducing the size of magnets (squares of 10x10x10 mm): a) 3D structure of magnets, b) Magnetic field line map for lines with 100 (dark green), 200 (light green), 300 (yellow), 400 (orange) and 500 (red) Gauss.

In this case, by using the ferromagnetic material only on the inside of the external magnet ring (Figure 4 and 5.a)), then the field becomes uniform also only on the external half of the channel (Figure 5.a)) and if we use this ferromagnetic material also on the external face of the internal magnet ring then the field will be uniform along the whole channel (Figure 5.b)). This uniformity of the magnetic field is extremely important for the correct generation of the Hall current inside the channel of the thruster and the subsequent generation of force.

(a) (b)

Figure 4. Disassembled PHALL II-B with partial ferromagnetic structure for magnets: a) perspective view, b) upper view.

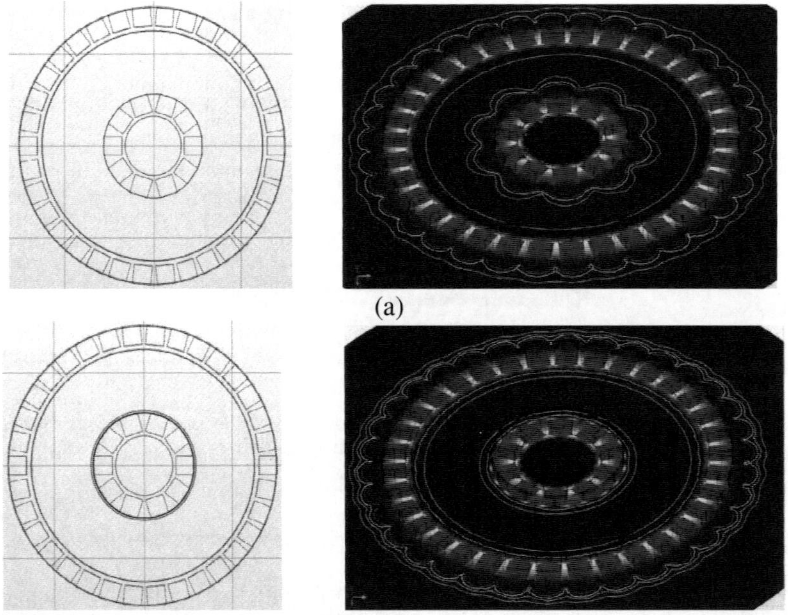

(a)

(b)

Figure 5. 3D structure of magnets and ferromagnetic material with magnetic field line map (100 – green, 200 – yellow, and 300 – red, Gauss) for the PHALL II-B magnetic field configuration with the **magnetic field perpendicular to the vertical wall of the channel** (square magnets of 10x10x10 mm) with ferromagnetic material: a) only on the inside of the external ring, b) plus the outside of the internal ring.

In experiments, we have observed directly the influence of magnetic field line uniformity in the Hall current generated plasma discharge. When the magnetic field lines are uniform only on the external half of the channel, then the Hall current only appears in that partial space (Figure 6.a)), but when all the channel has uniform magnetic field lines then the plasma of Hall current occupies all the available channel space (Figure 6.b)) increasing the generated propulsion force and efficiency.

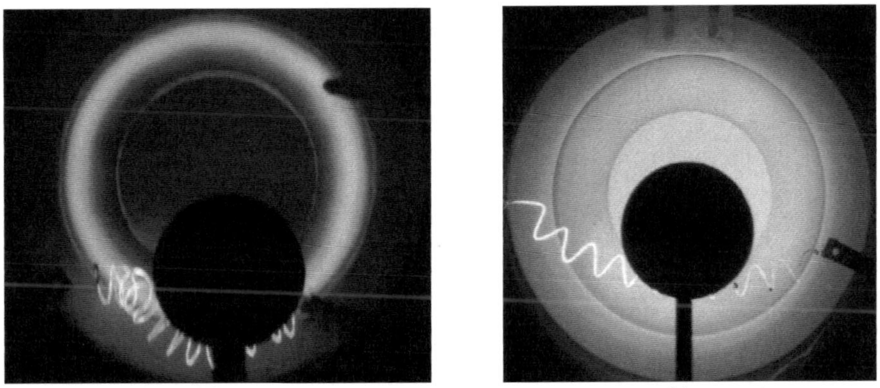

a) b)

Figure 6. Generated Hall current when: a) magnetic field lines are uniform only on the external half of the channel, and b) magnetic field lines are uniform across the whole channel.

3. Magnetic Field Simulation Results for B Field Parallel to Wall

As we mentioned, the use of this new magnetic field configuration using magnets (10x10x10 mm) has the additional advantage of them being able to be used either in the normal magnetic field configuration (Figures 1, 3 and 5) with the field perpendicular to the wall, or in a second configuration with the field parallel to the wall (Figure 7), known as "magnetic shielding" configuration, and avoiding in this way the strong erosion processes that occurs in the normal magnetic field configuration, significantly increasing the lifetime of the thruster. In this case, also in accordance with the results of the previous section, the magnetic field inside the channel becomes completely uniform when we use the ferromagnetic material both on the inside of the external ring and on the outside of the internal ring (Figure 7.b)).

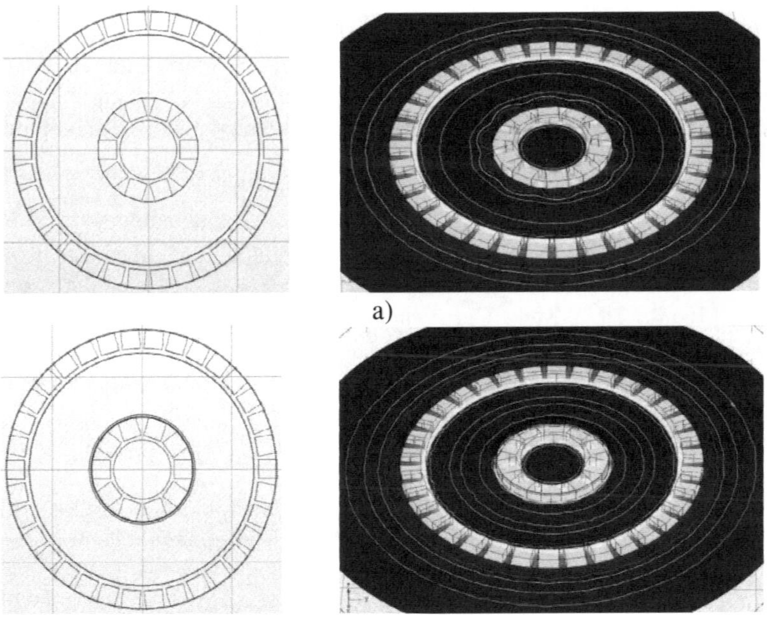

a)

b)

Figure 7.3D structure of magnets and ferromagnetic material with magnetic field line map (100 – green, 200 – yellow, and 300 – red, Gauss) for the PHALL II-B magnetic field configuration with the **magnetic field parallel to the vertical wall** ("**magnetic shielding**") of the channel (square magnets of 10x10x10 mm) with ferromagnetic material: a) only on the inside of the external ring, b) plus the outside of the internal ring.

Until the present moment, experiments have been made only with the magnetic field perpendicular to the wall (Figure 6.b)) with powers up to 800 W. A representative performance of our thruster at 450 W generates 36.4 mN of thrust with 2247 s of specific impulse. In the near future experiments will be performed also in the "magnetic shielding" configuration.

4. Conclusions

Initially, the intensity of the magnetic field at the center of the channel of PHALL II-A, where the discharge occurs was around 700-900 Gauss with 35.5 mm height magnets, producing a very strong magnetization of both ions and electrons (which inhibited the production of any propulsive force) generating a plasma torus. This problem was solved by replacing the long magnets (10x10x35.5 mm) by other smaller and shorter magnets (10x10x10 mm) and replacing the initial Aluminum support for the magnets by a different support with a ferromagnetic nature, which allows for a strong reduction of the magnetic field intensity in the propulsion channel to levels where only electrons are magnetized (100-300 Gauss) on one side, and on the other side it also allows for a much better magnetic field line uniformity inside the channel. In this case, by using the ferromagnetic material only on the outside ring, then the field becomes uniform only on the external half of the channel. But if we use this ferromagnetic material on both theinside and outside magnet rings, then the magnetic field becomes uniform in the whole channel. This uniformity of the magnetic field is extremely important for the correct generation of the Hall current in the thruster channel and consequently important for the generation of thrust.

The implantation of this new magnetic field configuration has the additional advantage of these magnets being able to be used both for the usual magnetic field configuration with the magnetic field perpendicular to the walls, and also able to be placed in a second different configuration with the magnetic fields directed parallel to the walls, known as "magnetic shielding" configuration, which is able to avoid by at least three orders of magnitude the strong erosion of the walls associated with Hall thruster operation and increasing in a very significant way the lifetime of these thrusters. These aspects will be tested in the near future.

Acknowledgements

The main author (A.A.M.) gratefully acknowledges the Brazilian Space Agency (AEB), the National Institute for Space Research (INPE) and the National Council of Technological and Scientific Development (CNPq) for the attribution of a Post-Doc Scholarship for the development of electric propulsion systems to use in satellites.

5. References

[1] Ferreira J L, Possa G C, Moraes B S, Souza J H C, Carvalho G M, Mourão D C and Winter O C 2009 *Advances in Space Research* Vol. 1 pp 166

[2] Winter O C, Mourão D C, Melo C F, Macau E N, Ferreira J L and Carvalho J P S 2010 *Mathematical Problems in Engineering/Space Dynamics* Vol. 2009 article ID 159287 pp.1

[3] Oliveira T C, Rocco E M, Ferreira J L and Prado A F B A 2013 *Mathematical Problems in Engineering/ Space Dynamics*, Vol. 2013, article ID 945030, pp. 1

[4] Moraes B S, Ferreira J L, Ferreira I S, Winter O C and Mourão D C 2014 *Journal of Physics: Conference Series* 511 012074

[5] Ferreira J L, Martins A A, Miranda R, Schelin A, Alves L S, Costa E G, Coelho H O, Serra A C B and Nathan F 2016 *Computer and Applied Mathematics/Orbital Dynamics* Vol. 35 (3) pp 711

[6] Goebel D M and Katz I 2008*Fundamentals of Electric Propulsion: Ion and Hall Thrusters* (JPL Space Science and Technology Series, Jet Propulsion Laboratory, California Institute of Technology)

IOP Conf. Series: Journal of Physics: Conf. Series **911** (2017) 012025 doi:10.1088/1742-6596/911/1/012025

Rendezvous and Quasi-Rendezvous Maneuvers With Space Debris

T C F Carvalho[1], A D C Jesus[1], L S Ferreira[1], R R Sousa[2]

[1] UEFS, Universidade Estadual de Feira de Santana, Feira de Santana - BA, Brasil.
[2] UNESP, Universidade Estadual Paulista Júlio Mesquita Filho. Guaratinguetá,SP. Brasil.

E-mail: Tham_mia@hotmail.com

Abstract. Rendezvous maneuvers are very useful in space missions operations. The encounters between spacecraft, between a spacecraft and a space debris, between a probe and a celestial body, between a spacecraft and a nearby object (NEO) as a mitigation measure to protect the Earth against collision and other applications, make use of these maneuvers. We define Quasi-Rendezvous maneuvers as those performed at low velocity, such that damage to the spacecraft is negligible. In this work, we study the distribution of these maneuvers as a function of the final relative velocities between a spacecraft and a spatial debris. The results were divided into two types: the first, for the dynamics subject only to the terrestrial gravitational force and the second, including the propulsion force acting on the vehicle. They show that the maneuvers propelled can approach the objects, overcoming the effect of the gravity of the earth for very small speeds.

1. Introduction

The Rendezvous between the two space objects must occur without collisions between them, that is, the relative position and velocity must be null in same time. This maneuver is one completely constrained, with several applications, mainly for the space station and space debris. Currently this maneuver has been designed for the strategies of mitigation of debris and the risk of collision of bodies near the Earth in 1960 [1]. After him, many authors studied this maneuver under many conditions and constraints. Stern [2] in 1984 approximated the Clohessy-Wiltshire equations to the small time transfer and obtained non-jouned rectilinear trajectories. The generalization to planar, minimum consumption Rendezvous case, in the general central force field was done by Humi [3] in 1993. In this year Abramovitz and Grunwald [4] developed an iterative graphical method to the optimal and planar Rendezvous inside many spacecraft of one space station environment, under several operational constraints saving more than 30 per cent fuel. In 1995, Yu [5] showed that an stable equilibrium state can occur in the relative motion between two close spacecrafts to Rendezvous inside a local coordinate system. Prado [6] also in 1995, derived an algorithm to solve optimal Rendezvous maneuvers with two impulses for a mono-revolution transfer or a multi-revolutions transfer, coplanar or non-coplanar. He found fits of the fuel consumption as function of transfer time. In 2001 Prado and Felipe [7] used impulsive control to study the Rendezvous maneuvers. A chaser-target rendezvous problem was solved by Carpenter [8] using a genetic algorithms. They used the Clohessy-Wiltshire equations as a linear approximation for preliminary mission planning. Another chaser-target type problem is studied by Kim and Spencer [9] with minimum fuel consumption as the objective function. Olsen and Fowler [10] also adopted the genetic algorithms to generate a near optimal solution to a rendezvous problem using elliptic orbits. Crispin [11],[12] obtained solutions to rendezvous problems as nonlinear discrete or continuous time optimal control problems with terminal

IOP Conf. Series: Journal of Physics: Conf. Series **911** (2017) 012025 doi:10.1088/1742-6596/911/1/012025

constraints. We studied Rendezvous maneuvers non-ideal, considering propulsion thrust deviations and mass variation [13]. We find the relative final position deviations w.r.t. the direction angles deviations, penalized by functions derived from the mass variation.

In 2015, we use the Rendezvous strategy for collision study of a space vehicle with a cloud of space debris [14]. In 2014 studies on the cloud of debris resonant with the earth's rotation angular velocity [15] and the modeling of the cloud of debris as fluid in an equation of continuity [16] were performed. The results obtained showed resonant movements for long periods, others irregular and an analytical expression for the evolution in the time of the density of the cloud. A cloud of debris spread by the interaction with the gravitational field and the obtaining of new trajectories of each debris belonging to the cloud by "Patched-Conics" can be found in [17]. This present study includes the possibility of Quasi-Rendezvous and our propelled maneuvers are continuous. Our approach is not optimal, but we use the propulsion system to control the terminal velocity of nearby objects in a Rendezvous or Quasi-Rendezvous maneuver around Earth.

2. Mathematical model

The dynamics between two space objects in the earth's gravitational field can be studied, using the relative motion equations found by Clohessy-Witshire [1]. These authors established the initial conditions in relative position and velocity favorable to a Rendezvous between two spacecraft. Jesus et al [18] included the propulsion force in the relative dynamics between two space objects and found a solution that can be used to perform an evasive maneuver against a collision with space debris. In this work, we will use the model of Jesus et al [18] to study the maneuvers of Rendezvous and Quasi-Rendezvous between a spacecraft and space debris. The coordinates of the relative velocity between these objects, subject to the earth's gravitational field, are:

$$\dot{x}_{(t)} = \dot{x}_o cos(\omega t) + (2\dot{y}_o + 3\omega x_o) sin(\omega t) \tag{1}$$

$$\dot{y}_{(t)} = -2\dot{x}_o sin(\omega t) - (4\dot{y}_o + 6\omega x_o) cos(\omega t) - (3y_o + 6\omega x_o) \tag{2}$$

$$\dot{z}_{(t)} = \dot{z}_o cos(\omega t) - z_o \omega sin(\omega t) \tag{3}$$

This solution was obtained for the situation of objects near each other with respect to the distance to the center of the Earth. The reference system is centered on the space vehicle that has circular orbit around the planet with angular velocity,$\vec{\omega} = \omega\hat{k}$, and the y-axis is in the direction radius vector r that links the objects. With these equations we can study the Rendezvous condition (r = 0 and v = 0) and the quasi-Rendezvous condition (r = 0, v ~ 0). These conditions can be obtained naturally in the gravitational field of the Earth. Nevertheless, a propelled maneuver can increase and control the possibilities of occurrence of these trajectories. This control will allow more safety and precision in the approximation of space objects. The solution of the relative dynamic equations with propulsion force was obtained by Jesus et al [18], whose Cartesian coordinates for the final relative velocity are:

$$\dot{x}(t) = 2Awcos(wt) + 2Bwsen(wt) - \textstyle\sum_{n=1}^{\infty} n\gamma F_n e^{-n\gamma t} + E \tag{4}$$

$$\dot{y}(t) = Bwcos(wt) - Awsen(wt) - \textstyle\sum_{n=1}^{\infty} n\gamma C_n e^{-n\gamma t} \tag{5}$$

$$\dot{z}(t) = Iwcos(wt) - Hwsen(wt) + \textstyle\sum_{n=1}^{\infty} n\gamma J_n e^{-n\gamma t} \tag{6}$$

IOP Conf. Series: Journal of Physics: Conf. Series **911** (2017) 012025 doi:10.1088/1742-6596/911/1/012025

This dynamics obtained for a propulsion system, assuming exponential mass variation of the propellant. All coefficients depend on the relative initial conditions,$(\vec{r}_o, \dot{\vec{r}}_o)$ and technological parameters characterize the propulsion system. These parameters are: $\vec{v_e}$- gas exhaust velocity; $\chi > 1$ - mass factor, ie, the ratio of spacecraft mass (M_o) to initial propellant mass (m_o) and; $\gamma > 0$ - motor power factor. With this model it is possible to control the approximation of the objects (the final values of $x_{(t)}, y_{(t)}$ and $z_{(t)}$) through the performance of the vehicle's propulsion system, establishing new Rendezvous and/or Quasi-Rendezvous conditions with propelled maneuvers. The coefficients in the Equantions $(4)-(6)$, in general, we can write them as $L = (\vec{r}_o, \dot{\vec{r}}_o, \vec{v_e}, \gamma, \chi)$. They are in the Appendix.

3. Numerical Simulation

In this section, we show the results of the simulations of the Rendezvous and Quasi-Rendezvous maneuvers, initially, under the exclusive action of the Earth's gravitational force and then for the maneuvers under the action of the propulsion force of the space vehicle. The Rendezvous condition is difficult to establish from the technical point of view if we consider infinite precision for the final relative position and velocity between the objects. The Quasi-Rendezvous condition is more likely, since depending on the range of small velocity we desire, the propulsion system can adjust the dynamics of the vehicle to obtain it. However, such propulsion systems are also limited and fail to implement any small speed ranges.

3.1. Natural Rendezvous/Quasi-Rendezvous

In the simulations we used time Rendezvous (or collision time) equal to 3,000s. This time is sufficient for the processing operations of the vehicle's internal computer to calculate and implement the maneuver, escaping from the debris or coupling to it. The distributions of the Quasi-Rendezvous possibilities are shown in Figures 1(a), (b) and 2, below.

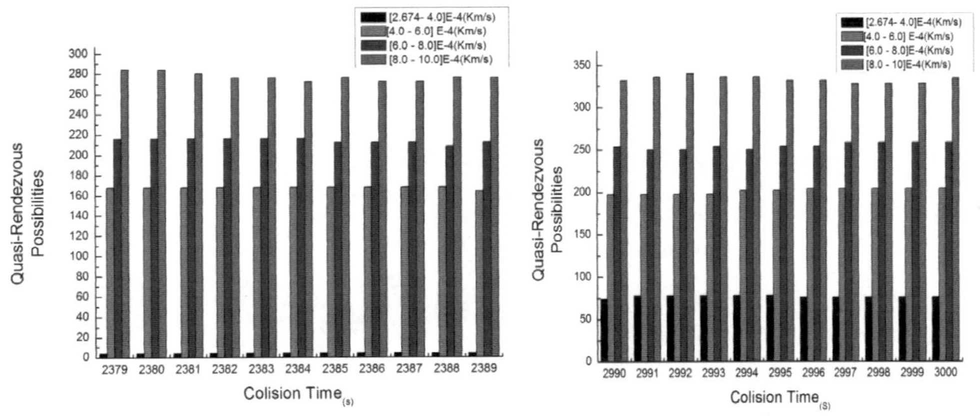

Figure 1- Quasi-Rendezvous Possibilities vs. Time Collision, $v \approx 10^{-4}$ km/s: (a) [2,379 to 2,389]s, (b) [2,990 to 3,000]s

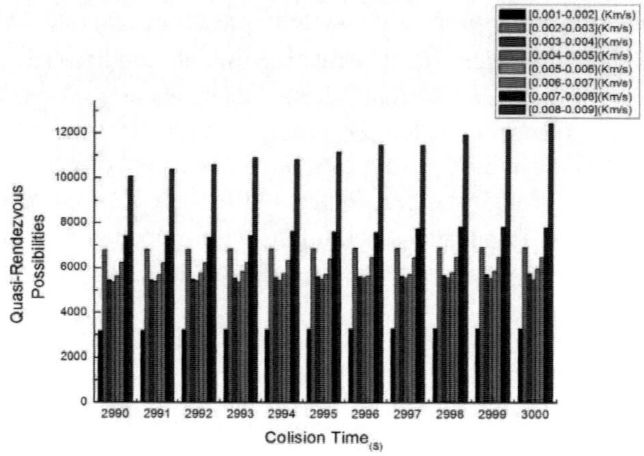

Figure 2- Quasi-Rendezvous Possibilities vs. Time Collision,_ $v \approx 10^{-3}$ km/s: [2,990 to 3,000]s

These results show that the Quasi-Rendezvous (or Rendezvous approximate) possibilities increases with time maneuver and for velocities also increasing. That is, when objects are subject to the gravitational field, they tend to meet naturally, if their final relative velocities are increasing. Thus, the Rendezvous possibilities decrease significantly with increasing time. Obviously, bodies accelerated by the gravitational field over time of exposure to it. The results also show that depending on the accuracy required, a Quasi-Rendezvous be accepted as a Rendezvous between objects. Figure 3, below, shows the distribution of possibilities as a function of the final relative velocity between them.

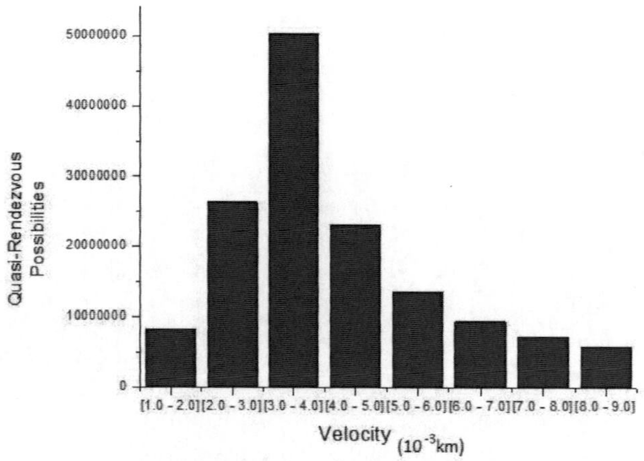

Figure 3- Quasi-Rendezvous Possibilities vs. Final Relative Velocity

We observe a very large number of approximations between objects with small but not null velocities. There is a peak of possibilities in the range of $[3{,}0 - 4{,}0]x10^{-3} km/s$ and the distribution seems to obey an almost Gaussian law. Thus, with increasing velocity, the bodies have enough energy to escape from the Rendezvous condition, and therefore from the Quasi-Rendezvous.

3.2. Propelled Rendezvous/Quasi-Rendezvous

In this section we show the results for the case of Rendezvous and/or Quasi-Rendezvous maneuvers using the propulsion system. We used technological parameters by Jesus et al [18], ie, $\chi = 10, \gamma = 10^{-6}\, 1/s, v_e = 2.5\, Km/s$. Table 1 below compares results of natural and propelled maneuvers. We observed that, in general, the natural maneuvers present more possibilities of Quasi-Rendezvous in relation to those propelled. As the velocity increases this trend is characterized. However, the propelled maneuvers allow Quasi-Rendezvous to velocities in $[10^{-5} - 10^{-4}]$ km/s, not occurring the natural ones. We understand that the propulsion controls the maneuvers, such that favor the Rendezvous (precision:10^{-5}) that could not occur naturally under the terrestrial gravity. For higher velocities, propulsion reduces the possibilities of Rendezvous and Quasi-Rendezvous. Figures 1(a), (b), below show these results.

Table 1 – Possibilities Natural and Propelled Quasi-Rendezvous

Maneuver	[10e-6–10e-5] (km/s)	[10e-5 – 10e-4] (km/s)	[10e-4 – 10e-3] (km/s)	[10e-3 – 10e-2] (km/s)	[10e-2 – 10e-1] (km/s)
Natural	0.000	0.000	1046870.000	149326716.000	40674360.000
Propelled	0.000	851.000	895675.000	100653569.000	25835985.000

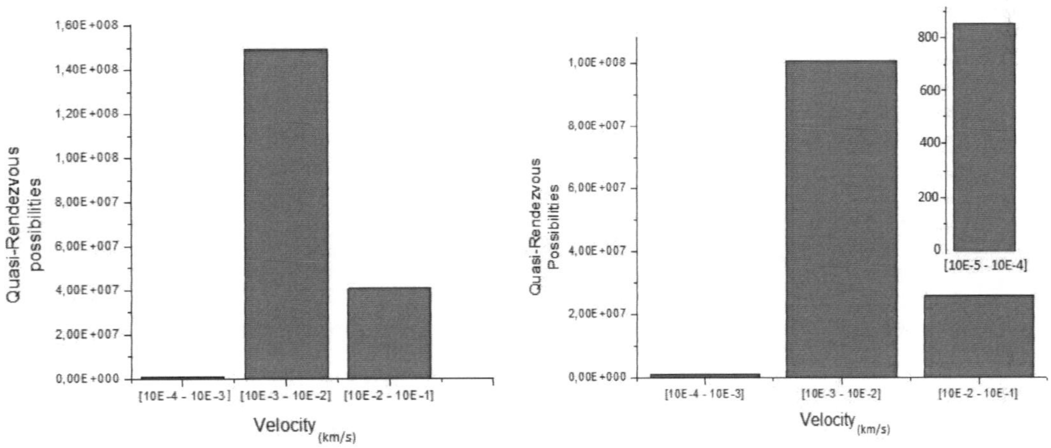

Figure 4 - Quasi-Rendezvous Possibilities vs. Final Relative Velocity: (a) Natural;(b) Propelled Maneuvers

4. Conclusions

We studied Rendezvous and Quasi-Rendezvous natural maneuvers (subject only to the gravitational field) and propelled between a spacecraft and a space debris. Rendezvous maneuvers are rare, but are made possible by propulsion if we consider zero equal to 10^{-5}. This accuracy can not be achieved naturally. In addition, propulsion reduces the possibility of Rendezvous for high velocities. Quasi-Rendezvous is more frequent at higher velocities if bodies are exposed to the gravitational field for a long time. This result also shows that in the space environment debris occurs with very small relative velocities different from those expected in LEO (7.76 km/s). Of course, this apparent discrepancy is explained by the fact that our model does not include other nongravitational forces acting. In this environment, which would interfere in the distribution of the possibilities of encounters between objects.

5. Appendice A – Equations Coefficients

$$A=\left\{\frac{2\dot{x}_o}{w}-3y_o+\frac{2v_{ex}}{w}\ln\left(\frac{\chi+1}{\chi}\right)-\sum_{n=1}^{\infty}\frac{(-1)^{n+1}}{n\chi^n}\left(\frac{2v_{ex}}{w}+\frac{n\gamma v_{ey}}{w^2}\right)\frac{1}{1+\left(\frac{n\gamma}{w}\right)^2}\right\} \quad (A.1)$$

$$B=\left\{\frac{\dot{y}_o}{w}+\frac{v_{ey}}{w}\ln\left(\frac{\chi+1}{\chi}\right)+\sum_{n=1}^{\infty}\frac{(-1)^{n+1}}{n\chi^n}\left(\frac{2n\gamma v_{ex}}{w^2}+\frac{v_{ey}}{w}\right)\frac{1}{1+\left(\frac{n\gamma}{w}\right)^2}\right\} \quad (A.2)$$

$$E=\left\{6wy_o-3\dot{x}_o-3v_{ex}\ln\left(\frac{\chi+1}{\chi}\right)\right\} \quad (A.3)$$

$$F_n=\frac{(-1)^{n+1}}{n\chi^n}\left(\frac{4v_{ex}}{n\gamma}+\frac{2v_{ey}}{w}\right)\frac{1}{1+\left(\frac{n\gamma}{w}\right)^2}-\frac{v_{ex}}{n\gamma} \quad (A.4)$$

$$G=x_o+\frac{2\dot{y}_o}{w}+\frac{2v_{ey}}{w}\ln\left(\frac{\chi+1}{\chi}\right)+\sum_{n=1}^{\infty}\frac{(-1)^{n+1}}{n^2\chi^n}\cdot\frac{3v_{ex}}{w} \quad (A.5)$$

$$C_n=\frac{(-1)^{n+1}}{n\chi^n}\left(\frac{2v_{ex}}{2}+\frac{n\gamma v_{ey}}{w^2}\right)\frac{1}{1+\left(\frac{n\gamma}{w}\right)^2} \quad (A.6)$$

$$D=4y_o-\frac{2\dot{x}_o}{w}-\frac{2v_{ex}}{w}\ln\left(\frac{\chi+1}{\chi}\right) \quad (A.7)$$

$$H=z_o+\sum_{n=1}^{\infty}\frac{(-1)^{n+1}v_{ez}\gamma}{\chi^n w^2}\frac{1}{1+\left(\frac{n\gamma}{w}\right)^2} \quad (A.8)$$

$$I=\frac{\dot{z}_o}{w}-\frac{v_{ez}}{w}\ln\left(\frac{\chi+1}{\chi}\right)+\sum_{n=1}^{\infty}\frac{(-1)^{n+1}}{n\chi^n w}v_{ez}\frac{1}{1+\left(\frac{n\gamma}{w}\right)^2} \quad (A.9)$$

$$J_n=\sum_{n=1}^{\infty}\frac{(-1)^{n+1}}{n\chi^n w}v_{ez}\frac{1}{1+\left(\frac{n\gamma}{w}\right)^2} \quad (A.10)$$

6 . References

[1] Clohessy W H and Wiltshire R S 1960 Terminal guidance system for satellite rendezvous. Journal of Aerospace Science 27(9) 653–674.
[2] Stern S A A 1984 Rectilinear guidance strategy for short orbital transfers. *Journal Spacecraft and Rockets* **21(6)** 542–545.
[3] Humi M 1993 Fuel-optimal rendezvous in a general central force field. *Journal of Guidance, Control, and Dynamics* **16(1)** 215–217.
[4] Abromovitz A and Grunwald A 1993 Interactive method for planning constrained, fuel-optimal orbital proximity operations. *American Astronautical Society* 93-309.

[5] Yu S 1993 Terminal spacecraft coplanar rendezvous control *Journal of Guidance, Control, and Dynamics* **18(4)** 838–842.

[6] Prado A F B A 1995 Optimal Rendezvous maneuvers for space vehicles. Proceedings, CDRoom, *XIII Brasilian Conference of Mechanics Engineering,* 12-15 December.

[7] Prado A F B A and Felipe G 2001 Manobras de Rendezvous entre orbitas Keplerianas com Controle Impulsivo. *SBA Controle e Automao* **12(2)** 156–162.

[8] Carpenter B & Jackson B 2003 Stochastic optimization of spacecraft rendezvous trajectories, *Advances in the Astronautical Sciences 113:* 219–232.

[9] Kim Y H & Spencer D B 2002 Optimal spacecraft rendezvous using genetic algorithms, *Journal of Spacecraft and Rockets* **39(6):** 859–865.

[10] Olsen C & Fowler W 2005 Characterization of the relative motions of rendezvous between vehicles in proximate, highly elliptic orbits, *Advances in the Astronautical Seciences* **119:** 879–895.

[11] Crispin Y 2006 An Evolutionary Approach to Nonlinear Discrete-Time Optimal Control With Terminal Constraints, Informatics in Control, Automation and Robotics I, Springer, Dordrecht, Netherlands.

[12] Crispin Y 2007 Evolutionary Computation for Discrete and Continuous Time Optimal Control Problems, Informatics in Control, Automation and Robotics II, Springer, Dordrecht, Netherlands.

[13] Jesus A D C Teles T N 2007 Rendezvous Maneuvers under Thrust Deviations and Mass Variation. *Nonlinear Dynamics and Systems Theory* **7(3)** 279-288.

[14] Jesus A D C Sousa R R. Neto E V 2015 Evasive Maneuvers in Route Collision with Space Debris Cloud. *Journal of Physics: Conference Series* **641** 012021.

[15] Sampaio J C Wnuk E Vilhena de Moraes R and Fernandes S S 2014 Resonant Orbital Dynamics in LEO Region: Space Debris in Focus *Mathematical Problems in Engineering* **v.2014**.

[16] Letizia F Colombo C Lewis HG McInnes CR 2013 Debris cloud evolution in Low Earth Orbit. *In: 64th International Astronautical Congress. International Astronautical Federation.* IAC-13.A6 pp12

[17] Formiga J K S Gomes V M. & de Moraes R V 2017 Orbital effects in a cloud of space debris making a close approach with the earth *Comp. Appl. Math* ***v*.5**.

[18] Jesus A D C et al 2012 Evasive Maneuvers in Space Debris Environment and Technological Parameters *Mathematical Problems in Engineerin*, Hindawi Publishing Corporation, **v2012,** pp15.

IOP Conf. Series: Journal of Physics: Conf. Series **911** (2017) 012026 doi:10.1088/1742-6596/911/1/012026

Numerical analysis of orbital transfers to Mars using solar sails and attitude control

M C Pereira[1], C F de Melo[1], L G Meireles[2]

[1] Federal University of Minas Gerais, UFMG, Belo Horizonte (MG), Brazil
[2] National Institute for Space Research, INPE, São José dos Campos (SP), Brazil

cristiano.fiorilo@demec.ufmg.br

Abstract. Solar sails present a promising alternative method of propulsion for the coming phases of the space exploration. With the recent advances in materials engineering, the construction of lighter and more resistant materials capable of impelling spaceships with the use of solar radiation pressure has become increasingly viable technologically and economically. The studies, simulations and analysis of orbital transfers from Earth to Mars proposed in this work were implemented considering the use of a flat solar sail. Maneuvers considering the delivery of a sailcraft from a Low Earth Orbit to the border of the Earth's sphere of influence and interplanetary trajectories to Mars were investigated. A set of simulations were implemented varying the attitude of the sail relative to the Sun. Results show that a sailcraft can carry out transfers with final velocity with respect to Mars smaller than the interplanetary Patched-conic approximation, although this requires a longer time of transfers, provided the attitude of the sailcraft relative to the Sun can be controlled in some points of the trajectories.

1.Introduction

Even though the concept of solar sails was presented in the 60's, material sciences was not developed enough to make sailcrafts available back then. The first successful launch only took place in 2010, when IKAROS mission (JAXA) started its operation [1] and opened the space sailing era. Other well-known initiatives are the Solar Sail Demonstrator, or Sunjammer, from NASA [2], and the LightSail initiative proposed by the Planetary Society [3]. The Sunjammer project was concluded in 2014, prior to flight testing, with interesting results. LightSail initiative is meant to be attached to a 3U CubeSat. LightSail I was launched in 2015 with the objective of testing the deployment of the sail, while in 2017, LightSail II is to be launched and fully tested.

Two objectives are proposed in this paper. The first one is to study the physical concepts and mathematical models for solar sails in order to implement numerical algorithms to simulate orbital transfers with such type of propulsion system. Secondly, to implement these simulations considering transfers from Earth to Mars with attitude control.

Following [4], the mathematical model considers the lightness vector **L** to be the thrust acceleration normalized to the local solar gravitational acceleration and resolved in the heliocentric inertial reference frame. The magnitude of the lightness vector can be defined by

$$L = \left(\frac{1}{2}\frac{\sigma_c}{\sigma}\right) n_x\left[\left(2r_{spec} + n_x\chi_f r_{diff} + \kappa a_f\right)n + \left(a_f + r_{diff}\right)u\right]. \quad (1)$$

Where σ_c is the critical sail loading and σ is the sailcraft sail loading. r_{spec} is the sailfront-side specular reflectance, r_{diff} is the sailfront-side diffuse reflectance, χ_f is the sailfront-side emission/diffusion coefficient, κ is the emission/diffusion net thrust dimensionless factor, a_f is the sailfront-side total absorptance. $\boldsymbol{n}$ is the sail's orientation unit vector, normal to its surface, and $\boldsymbol{u}$ is the direction of the incident sunlight on the sail, both resolved in the sailcraft orbital frame $x_s\, y_s\, z_s$. Two angles are important to define the orientation and the attitude of the sail in this frame, they are the azimuth α (in the orbital plane) and elevation δ (in the plane perpendicular to the orbital plane) – these quantities are shown in Figure 1. The critical sail loading is a constant defined by

$$\sigma_c \equiv 2\frac{I_{1AU}}{c g_{1AU}} \cong 1.5368\frac{g}{m^2}, (I_{1UA} = 1366\frac{W}{m^2}), \tag{2}$$

and where $g_{1AU} = 5.930\times10^{-3}$ m/s² is the gravitational acceleration at 1 AU from the Sun, and c = 299792458 m/s denotes the speed of light. The sailcraft sail loading is determined by

$$\sigma \equiv \frac{m}{S}, \tag{3}$$

where m is the sail's mass and S is the sail's surface area. Finally, the emission/diffusion net thrust dimensionless factor is defined by

$$\kappa \equiv \frac{\chi_f \epsilon_f(T) - \chi_b \epsilon_b(T)}{\epsilon_f(T) + \epsilon_b(T)}, \tag{4}$$

where T is the sail's temperature, χ is the emission/diffusion coefficient and ϵ is the emittance as a function of T. The subscripts f and b indicate whether the coefficient corresponds to the front-side or back-side of the sail, respectively. Let Ξ be the transformation matrix from the sailcraft orbital frame to the heliocentric inertial frame, which is defined by

$$\Xi = (\boldsymbol{r} \quad \boldsymbol{h} \times \boldsymbol{r} \quad \boldsymbol{h}), \tag{5}$$

where $\boldsymbol{r}$ is the outward radial direction and $\boldsymbol{h}$ is the direction of the sailcraft's orbital angular momentum per unit of mass. From equations (1) and (5), it is possible to calculate the sailcraft's acceleration resolved in the heliocentric inertial frame considering:

$$\boldsymbol{A}^{HIF} = \Xi g L, \tag{6}$$

where g is the local solar gravitational acceleration.

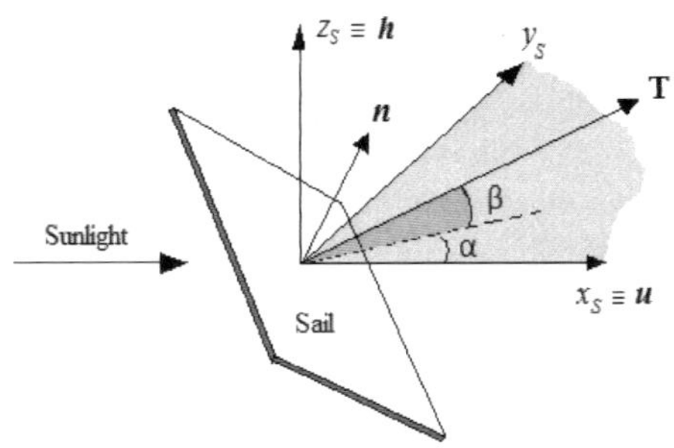

Figure 1. Scheme of the orientation in $x_S y_S z_S$ frame fixed in Sail. **T** represents the thrust on the sail due to the interaction with the solar radiation.

2. Mission parameters

The simulations were implemented considering as dynamical system the four-body problem Sun-Earth-Moon-sailcraft [5,6], with the Earth and the Moon in circular orbits. The acceleration components due to solar radiation pressure on the sailcraft are derived from the equation (5) and (6) and given by

$$\boldsymbol{A}^{HIF} = \frac{GM_{Sun}}{R^2}\begin{bmatrix} \frac{l_x}{R}x + \frac{l_y}{HR}\left[z(zv_x - xv_z) - y(xv_y - yv_x)\right] + \frac{l_z}{H}(yv_z - zv_y) \\ \frac{l_x}{R}y + \frac{l_y}{HR}\left[x(xv_y - yv_x) - z(yv_z - zv_y)\right] + \frac{l_z}{H}(zv_x - xv_z) \\ \frac{l_x}{R}z + \frac{l_y}{HR}\left[y(yv_z - zv_y) - x(zv_x - xv_z)\right] + \frac{l_z}{H}(xv_y - yv_x) \end{bmatrix}. \quad (7)$$

Where G is gravitational constant, M_{Sun} is the Sun's mass, R is the magnitude of the sailcraft's position in the inertial reference frame, fixed at the Sun's center of mass, (x, y, z) and (v_x, v_y, v_z) are the components of the sailcraft's position and velocity, respectively. H is the magnitude of the sailscraft's orbital angular momentum. Equation (8) indicates the components (l_x, l_y, l_z) of the lightness vector, derived from the equation (1) as a function of the angles α and δ.

$$\begin{cases} l_x = \left(\frac{1}{2}\frac{\sigma_c}{\sigma}\right)\left[2r_{spec}cos^3\delta\ cos^3\alpha + (\chi_f r_{diff} + \kappa a_f)cos^2\delta\ cos^2\alpha + (a_f + r_{diff})\cos\delta\ \cos\alpha\right] \\ l_y = \left(\frac{1}{2}\frac{\sigma_c}{\sigma}\right)\left[2r_{spec}cos^3\delta\ cos^2\alpha\ sin\alpha + (\chi_f r_{diff} + \kappa a_f)cos^2\delta\ cos\alpha\ sin\alpha\right] \\ l_z = \left(\frac{1}{2}\frac{\sigma_c}{\sigma}\right)\left[2r_{spec}cos^2\delta\ cos^2\alpha\ sen\delta + (\chi_f r_{diff} + \kappa a_f)cos\delta\ cos\alpha\ sin\delta\right] \end{cases}. \quad (8)$$

The values of the optical coefficients considered were taken from the values already achieved by the Jet Propulsion Laboratory (JPL/NASA) [7] and are presented in Table 1.

Table 1. Sail's optical coefficients [7].

	r	S	ϵ_f	ϵ_b	b_f	b_b
Ideal Sail	1	1	0	0	2/3	2/3
Square Sail	0.88	0.94	0.05	0.55	0.79	0.55
Heliogyro	0.88	0.94	0.05	0.55	0.79	0.55

The values were taken from the second line of the Table 1 for a square sail. The first column corresponds to the reflectance coefficient, whilst the second column is the specular reflection ratio. Nevertheless, a value of 90% specular reflection ratio was considered for the sailfront-side. By considering a null transmittance, the value of the absorptance can be defined from the value of the reflectance. The third and fourth columns are the front-side and back-side emittance values, respectively, while the fifth and sixth columns correspond to the front-side and back-side emission/diffusion coefficients.

Still, according to the JPL, current technologies allow the construction of a sail with sailcraft loading of 5.27 g/m² [7]. And taking LightSail 1 mission [3] into account, a mass of 20 kg was considered and, indirectly, a surface area of 3795 m². With that, all the coefficients and parameters of the sail were defined, making it possible to determine the acceleration in the inertial heliocentric frame at every iteration.

2.1 Constant attitude simulations

A total of 304 simulations were performed considering the following conditions:

- Eight initial positions (A, B, C, D, E, F, G, H) on the border of the Earth's sphere of influence, in the plane of the ecliptic, were considered. For a geocentric frame, these points have a separation of 45 degrees, as illustrated in Figure 2. To reach these points, impulsive Hohmann transfers from a Low Earth Orbits (LEO) were performed. During these maneuvers, the solar radiation pressure was considered not to act on the sail. Hohmann transfers proved to be more economic to start the mission from a LEO [6] in a scenario where the sail could be opened in a point on the Earth's sphere of influence. In terms of the magnitude of the total velocity change ($\Delta V_{H/\mathrm{Total}}$), it is required 3.157 km/s to deliver the sailcraft from a LEO to the Earth's sphere of influence through a Hohmann transfer.
- For each initial position, indicated in Fig. 2, a group of α values was considered {1°, 5°, 10°, 15°, 20°, 25°, 30°, 35°, 40°, 45°, 50°, 55°, 60°, 65°, 70°, 75°, 80°, 85°, 89°}, maintaining them constant along of each simulation;
- Since the orbits considered are coplanar, the δ value was always 0 degrees.

At the end of each simulation, the time of transfer and the position at the which the spacecraft inserted itself in the destination's planet sphere of influence were recorded. At this position, the destination planet's velocity was calculated and the difference between its value and the sailcraft's velocity (ΔV) was determined. Together with these information, it was also determined the destination planet's position relative to Earth at the beginning of the interplanetary transfer.

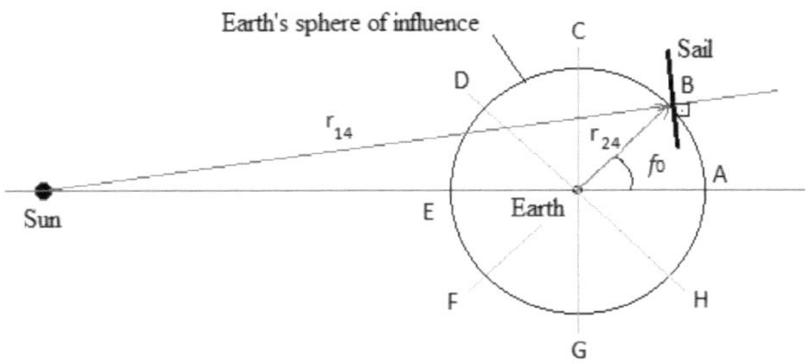

Figure 2. Relative position between the Sun, the Earth and the sailcraft. $\mathbf{r}_{14}$ and $\mathbf{r}_{24}$ are the position of the sailcraft relative to the Sun and the Earth, respectively. f_0 is the angular displacement (45 degrees) of the initial positions occupied by the sailcraft on the border of the Earth's sphere of influence (points: A, B, C, D, E, G, G, H).

2.2 Simulations with attitude control

To improve the results, especially regarding to the relative velocity between the sailcraft and the destination planet (ΔV) and the time of transfer, a procedure to change the attitude of the sailcraft has been implemented. This procedure consists in a few number of changes in the parameter α during the flight between the Earth's and the destination planet's sphere of influences. In general, as consequence, the solar radiation pressure will act to accelerate and then to deaccelerate the sailcraft reducing its velocity on arrival in the destination planet. However, other phases with acceleration and deacceleration can happen.

To consider the attitude change, the final state vector values of all the bodies (Sun, Earth, Moon and sailcraft) of the interrupted simulation were saved and inserted into a new simulation as the initial state vector of all the bodies. This new simulation would proceed with a different sail attitude.

Simulations starting from all positions on the Earth's sphere of influence (Fig. 2) were performed. The values of the attitude changes were chosen based on the results of the constant attitude simulations and the moment of the changes were evaluated considering the behavior of the value of the sailcraft's specific orbital energy.

3 Interplanetary transfers to Mars

The results achieved from the simulations were gathered in form of data plots and trajectory graphs are shown and discussed in this section.

3.1 Results for the transfers with constant attitudes α

For each transfer with α constant analyzed, the time of transfer and final relative velocity between the sailcraft and Mars (ΔV) were calculated, recorded and plotted. The behavior of a sailcraft's transfers to Mars is illustrated in Figure 3 where a clear relation between the time of transfer, ΔV and α can be seen for $f_o = 0$ degrees (Figure 2). The value of the constant attitude α that minimizes the time of transfer also maximizes the ΔV. The shortest time of transfer, 122.098 days, is found for $\alpha = 20$ degrees with $\Delta V = 12.296$ km/s. However, this is not acceptable for missions where the objective is, for example, to insert a space vehicle into an orbit around a planet. On the other hand, transfers with constant attitude α which offer low ΔV have higher times of transfer, as example, for $\alpha = 88$ degrees, the time of transfer is equal 14409.86 days (39.48 years) and $\Delta V = 0.473$ km/s, but this time of flight is also not acceptable.

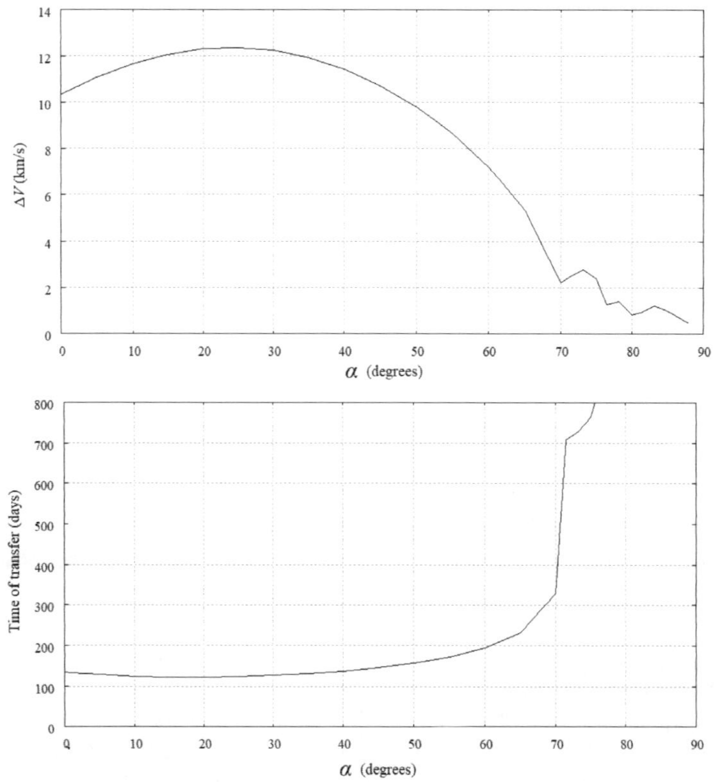

Figure 3. Transfer to Mars plot: time of transfer × α and $\Delta V \times \alpha$ for $f_0 = 0°$.

3.2 Earth-Mars transfer considering changes in the attitude α

An example of an Earth–Mars transfer can be observed in Figure 4. Four attitude changes of 5 degrees each are made during the transfer, the locations of the changes in the transfer trajectory are indicated by the little black dashes. According to Figure 3, if α is maintained constant and equal 60 degrees and for $f_0 = 0$ degrees, it is possible to transfer a sailcraft to Mars in 194 days, and the relative velocity between both is 7.210 km/s on arrival at a point on the Mars sphere of influence. Considering four attitude changes, it is possible to reduce this relative velocity to 2.090 km/s also at the same point on the Mars sphere of influence, but increasing the time of transfer to 305.87 days. The changes in the attitude are:

1) α = 60 degrees during the first 24.44 days.
2) First change α = 65 degrees from 24.44 days to 59.81 days, i. e., for 35.37 days.
3) Second change α = 70 degrees from 59.81 days to 103.71 days, i. e., for 43.9 days.
4) Third change α = 75 degrees from 103.71 day to 147.61 days, i. e., for 43.9 days.
5) Fourth change α = 80 degrees from 147.61 days to 305.87 days, i. e., for 158.26 days, when the sailcraft reaches the Mars sphere of influence.

In comparison with the interplanetary Patched-conic approximation [8], for which the minimal relative velocity between the Mars and a space vehicle found is 3.300 km/s at the same point and with a time of transfer of 177 days, in this example, the four changes in the angle α provide a reduction of 36% in the relative velocity between de Mars and the sailcraft against an increase of 129 days in its trip.

Even though this is not an optimal result essentially, but the strategy of changing α shows that the main obstacle to the use of sailcrafts in interplanetary exploration, that is, the constant increase in its velocity, can be solved if an onboard control system could provide a few minor changes in the α values during its journey.

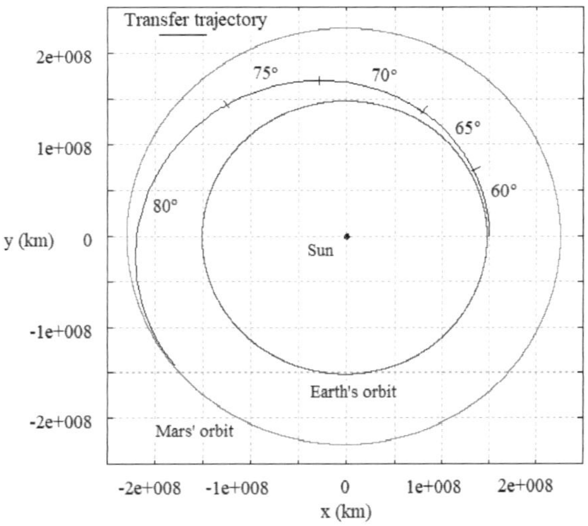

Figure 4. Transfer trajectory found considering four changes in the angle α of 5 degrees each.

4. Conclusion

The constant increase in the velocity can be considered as a problem to be overcome with regarding the use of sailcrafts in the planetary exploration. In this context, the possibility of changing the attitude of a sailcraft through few variations in the angle α during its journey across the space could reverse this constraint. This strategy allows to accelerate the spaceship at the beginning and reduce the relative velocity between the it the target celestial body at the end, although the time of transfer can increase. In the example presented here, in a transfer to Mars was possible to find a trajectory in which the relative velocity at the Mars sphere of influence is 2.090 km/s against 3.300 km/s found for a conventional impulsive transfer as determined by the interplanetary Patched-conic approximation.

Although it is important to note that these values were obtained without the use of any optimization algorithms, as they are the results of an introductory study on the application of solar sails on interplanetary transfers. Therefore, these are values with potential for refinement and improvement to reduce its dependency of chemical propulsion in the final trajectory adjustments.

IOP Conf. Series: Journal of Physics: Conf. Series **911** (2017) 012026 doi:10.1088/1742-6596/911/1/012026

References

[1] http://www.isas.jaxa.jp/en/missions/spacecraft/current/ikaros.html Last access: 26 Apr. 2017

[2] https://www.nasa.gov/mission_pages/tdm/solarsail/index.html Last access: 26 Apr. 2017

[3] The Planetary Society *LightSail: a solar sailing spacecraft from the Planetary Society Available* at: <http://sail.planetary.org/> Last access: 26 Apr. 2017.

[4] Vulpetti G, Johnson L and Matloff G L *Solar Sails: a novel approach to interplanetary travel* 2nd edition New York: Copernicus Books 2015 277 p.

[5] Murray, C.D. and Dermott, S.F. 1999 Solar System Dynamics, UK: Cambridge University Press.

[6] Meireles L G *Análise de Transferências Orbitais com uso de Velas Solares* Belo Horizonte: 2016.

[7] Mcinnes C R *Solar Sailing: Technology, Dynamics and Mission Applications*2nd edition London: Springer Praxis Books 2004 296 p.

[8] Bate, R.R.; Mueller, D.D.; White, J.E. Fundamentals of Astrodynamics. USA: Dover Publications Inc., 1971. 455p.

IOP Conf. Series: Journal of Physics: Conf. Series **911** (2017) 012027 doi:10.1088/1742-6596/911/1/012027

Orbital migration and Resonance Offset of the Kepler-25 and K2-24 systems

C. Charalambous[1], X. S. Ramos[1*], P. Benítez-Llambay[2], and C. Beaugé[1]

[1] Instituto de Astronomía Teórica y Experimental, Observatorio Astronómico, Universidad Nacional de Córdoba, Laprida 854, X5000BGR, Córdoba, Argentina

[2] Niels Bohr International Academy, The Niels Bohr Institute, Blegdamsvej 17, DK-2100, Copenhagen Ø, Denmark

E-mail: charalambous@oac.unc.edu.ar

Abstract. Based on the model described in [1], we present an analytical+numerical study of the resonance capture under Type-I migration for the Kepler-25 [2] and K2-24 [3] *Kepler* systems, both close to a 2/1 mean-motion resonance. We find that, depending on the flare index and the proximity to the central star, the average value of the period-ratio between two consecutive planets show a significant deviation with respect to the resonant nominal value, up to values well in agreement with the observations.

1. Introduction

Most of the planets found in *Kepler* multi-planetary systems lie outside mean-motion resonances (MMRs). This seems incompatible with a formation process strongly affected by planet migration and may either indicate that planets formed in-situ (e.g. [4–6]) or that planetary migration occurred in a turbulent environment, where resonance capture is not guaranteed (e.g. [7, 8]). However, the *Kepler* population also shows the existence of a statistically significant number of planetary pairs close to resonances, where the orbital period ratios is usually larger than the nominal value. It is generally believed that these systems are near-resonant and located outside the libration domain.

The origin of these near-resonant systems is a dilemma. Planetary migration in a turbulent disk [9] can lead to near-resonant configurations, although it is not clear whether this mechanism can reproduce the observed near-resonant distribution (e.g. [10]). This mechanism only seems to work for sub-Jovian bodies, and therefore, larger than the expected size for most of the *Kepler* systems.

In [1] we presented an analytical model that allow us to reproduce the general trend of the resonance offset, as the disk is assumed significantly flared and with a small scale height. This model strongly depends on the planetary masses, values which are not usually known for the *Kepler* systems, thus we performed Monte Carlo statistical analysis with no direct application to a given planetary system. Even so, the method has proved promising and able of reproducing the offset distribution around both the 2/1 and 3/2 MMRs, as well as predicting an increase in this value for planets closer to the central star.

From all the *Kepler* systems near the 2/1 MMR, just two cases have fairly credible estimations for the masses: Kepler-25 and K2-24, indicated as green and black circles in the left hand frame of Figure 1. Red circles correspond to systems detected by transits or Transit Time Variations (TTV). Bodies

* Present address: Niels Bohr International Academy, The Niels Bohr Institute, Blegdamsvej 17, DK-2100, Copenhagen Ø, Denmark

IOP Conf. Series: Journal of Physics: Conf. Series **911** (2017) 012027 doi:10.1088/1742-6596/911/1/012027

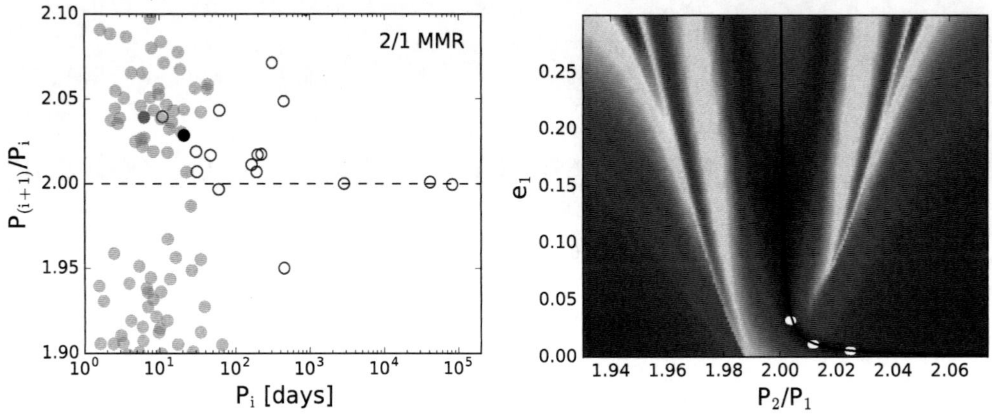

Figure 1. Left: Distribution of orbital period ratios, in the vicinity of the 2/1 resonance, as a function of the orbital period of the inner planet. Red circles identify planets detected by transits or TTV, while those discovered by other methods are depicted with open circles. The green circle indicates the location of Kepler-25 while the black one corresponds to K2-24. Data was obtained from `exoplanet.eu`. **Right:** Dynamical map of $max(\Delta e)$ for two-planet systems with $m_1 = 0.05 m_{\rm Jup}$ and $m_2 = 0.10 m_{\rm Jup}$ orbiting a central star of mass $m_0 = 1 M_\odot$, in the vicinity of the 2/1 MMR. The orbit of the outer planet was initially circular with $a_2 = 1$ AU, and all the angular variables where chosen equal to zero. The black continuous line marks the location of the zero-amplitude ACR-type librational solutions, estimated from the simple analytical model (eq. (4)). White dots are the result of three N-body simulations of resonance trapping.

discovered by any other method are identified by open circles. This distribution shows increasing $\Delta_{2/1}$ for planets closer to the star, indicating a possible smooth trend which, if confirmed, would indicate that the distribution found in different populations belong to the same functional form, and just differing from the distance to the star. It also shows little correlation with either the detection method or the stellar/planetary masses. Here, we apply the model of [1] to these two systems and attempt to constraint the properties of the protoplanetary disk that are consistent with their observed location and deviation from the exact resonance.

Right frame of Figure 1 shows a dynamical map for the 2/1 commensurability. We integrated series of two-planet systems with initial conditions in a grid defined in the $(P_2/P_1, e_1)$ plane and specifically chose $m_2/m_1 > 1$ to guarantee symmetric fixed points for the resonant angles (e.g. [11, 12]). The color code corresponds to the maximum value of $|e_1(t) - e_1(t = 0)|$ (denoted as $max(\Delta e)$) attained a during 10^3 years integration time. Darked (lighter) tones are associated to small (large) variations in the eccentricity of the inner planet. Although this indicator does not measure chaotic motion, it is an important tool to probe the structure of resonances and identify the locus of stationary solutions (so-called ACR solutions, see [11, 13]). It also helps to identify the separatrix delimiting the librational from the circulation domains (e.g. [14]). The black line shows the approximate location of the family of zero-amplitude ACR solutions characterized by the simultaneous libration of the both resonant angles.

2. The Ramos et al. Model

In the following we will assume two planets of masses m_1 and m_2 orbiting a star m_0, orbital periods $P_1 < P_2$ and in the vicinity of a first-order $(p+1)/p$ MMR. We define the *resonance offset* $\Delta_{(p+1)/p}$ as

$$\Delta_{(p+1)/p} = \frac{P_2}{P_1} - \frac{(p+1)}{p}, \tag{1}$$

whose value indicates the distance from the exact resonance.

Different values of $\Delta_{(p+1)/p}$ are attained in different parts of the disk. In order to study this, we use a resonant Hamiltonian neglecting secular perturbations to estimate the resonance offset as function of e_1, as well as a relation between the eccentricities of both planets:

$$\Delta_{(p+1)/p} = C_1(\alpha)\,\frac{m_2}{m_0}\frac{1}{e_1} \quad ; \quad e_2 = C_2(\alpha)\,\frac{m_1}{m_2}e_1, \tag{2}$$

(see [15, 16]), where the coefficients C_i depends solely on $\alpha = a_1/a_2$ ($C_1 \simeq 1.5$ and $C_2 \simeq 0.29$). For a given resonance, very low e_i are necessary to obtain a significant deviation from the exact resonance. However, since e_i does not attain zero for the ACR solution, the singularity at $e_1 = 0$ is never reached.

If the disk-driven planetary migration is sufficiently slow and smooth, we expect the orbital evolution to follow the pericentric branch into the librational domain and exhibit low-amplitude oscillations of the resonant angles. In such an ideal scenario, the final eccentricities and resonant offset $\Delta_{(p+1)/p}$ will depend on the relative strength between the eccentricity damping and orbital migration timescales ([11, 17]) τ_{e_i} and τ_{a_i}, respectively. Thus, the final outcome of a resonance trapping will depend on the ratios $\mathcal{K}_i = \tau_{a_i}/\tau_{e_i}$, which we denote as the *K-factors*.

We performed 3 N-body simulations including an ad-hoc external acceleration (e.g. [18]), set the values of τ_{a_i} at certain prefixed amounts, varied $\mathcal{K}_i$ and analyzed its effects on the resonance offset (white dots in the right frame of Figure 1). We chose planetary masses $m_1 = 0.05m_{\rm Jup}$ and $m_2 = 0.10m_{\rm Jup}$, and $1M_\odot$ for the central star. The outer planet was initially at $a_2 = 1$ AU, in circular orbit and all angles equal to zero. The initial conditions were integrated for 10^5 yrs. All values agree with the ACR loci given by expression (2) deduced from the analytical resonance model. Although these simulations indicate that is possible to obtain large values for the offset, they only appear attainable for K-factors of the order of 10^4, much higher than predicted by linear models of Type-I disk-planet migration (e.g. [18–20]). In [1], we found that it is possible to overcome this problem assuming a significant flare for the disk (of the order of $f \simeq 0.25$) in addition to a relatively small value for the disk aspect ratio ($H_0 \simeq 0.03$). This combination, in addition to moderately low values for the mass ratios of the planets, generates large deviations from exact resonance even with K-factors of the order of 10^2, well within the classical limits.

We assume a laminar disk with surface density $\Sigma(r) = \Sigma_0 r^{-\alpha}$ and aspect-ratio $H_r(r) = H_0 r^f$, where r is the distance to the central star in astronomical units. We will consider H_0, α and f as unknown parameters that will be estimated in accordance with the observed dynamical characteristics of the planetary systems.

Following [21] and [20], orbital migration and eccentricity damping timescales are approximated as

$$\tau_{a_i} = Q_a\frac{t_{{\rm wave}_i}}{H_{r_i}^2} \quad ; \quad \tau_{e_i} = Q_e\frac{t_{{\rm wave}_i}}{0.780} \quad ; \quad t_{{\rm wave}_i} = \frac{m_0}{m_i}\frac{m_0}{\Sigma(a_i)a_i^2}\frac{H_{r_i}^4}{\Omega(a_i)}. \tag{3}$$

In these expressions, H_{r_i} is the disk aspect-ratio in the position of each planet and $\Omega(a_i)$ their orbital frequency. Q_e is a constant introduced by [22] in order to reproduce the eccentricity damping rates from hydro-dynamical simulations, while $Q_a = Q_a(\alpha)$ is a function of the surface density profile. Finally, $t_{{\rm wave}_i}$ is the typical timescale of planetary migration.

Recently, the classical Type-I migration models were revised by [23], who considered the contribution of eccentricity damping to changes in the semimajor axis associated to (partial) conservation of the angular momentum. They found that the effective characteristic timescale for the orbital evolution should actually be given by $\tau_{a_{eff_i}} = \left(\tau_{a_i}^{-1} + 2\beta e_i^2\tau_{e_i}^{-1}\right)^{-1}$, where τ_{a_i} and τ_{e_i} maintain the same form as equations (3) and β is a factor that quantifies the fraction of the orbital angular momentum preserved during the migration. This modified migration timescale changes the K-factor, leading to a new "effective" form. This revised migration model, together with the analytical resonant Hamiltonian led in [1] to a relation between the disk properties and resonance offset in the form:

$$\Delta^2_{(p+1)/p} = \frac{2}{D}\left(C_1\frac{m_2}{m_0}\right)^2 \frac{\left[(1-D(1+\beta))\mathcal{K}_2\left(C_2\frac{m_1}{m_2}\right)^2 + (B+D\beta)\mathcal{K}_1\left(\frac{\tau_{a_2}}{\tau_{a_1}}\right)\right]}{1-\left(\frac{\tau_{a_2}}{\tau_{a_1}}\right)}, \tag{4}$$

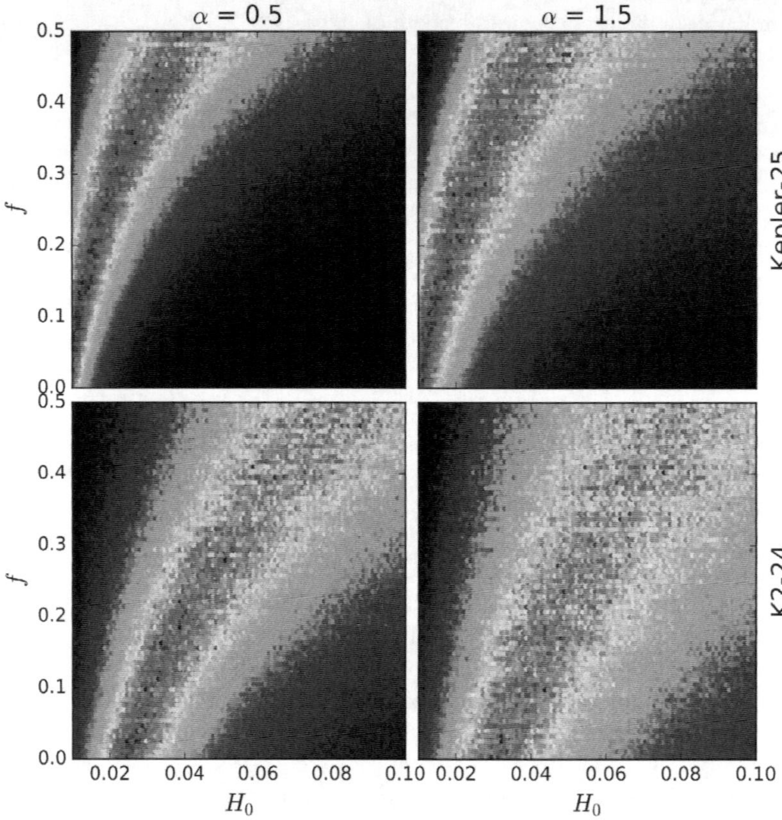

Figure 2. Results of Monte Carlo simulations for Kepler-25 (upper frames) and K2-24 (bottom panels) for different density slopes values. In the left we show $\alpha = 0.5$ and in the right $\alpha = 1.5$. Red colors indicate more possible pairs (H_0, f) for an observed $\Delta_{2/1}$, than the blue values.

where the parameters B and D depend both on planetary mass ratio and the resonance under consideration

$$D = \frac{1}{(p+1)}\left(1 + \frac{a_1}{a_2}\frac{m_2}{m_1}\right)^{-1} \quad ; \quad B = \frac{m_1 n_2 a_2}{m_2 n_1 a_1} + D. \tag{5}$$

The importance of expression (4) lies in the fact that it shows that large values of the offset may be obtained if the denominator is sufficiently small, independently of the K-factors. Since the ratio τ_{a_2}/τ_{a_1} depends on the planetary mass ratio (as well as on parameters f and α), it is possible to obtain values of $\Delta_{(p+1)/p}$ consistent with the observed planetary systems, as long as the parameters lie within certain values.

3. Application to Kepler-25 and K2-24 systems

Of 165 planetary pairs lying in the vicinity of the 2/1 resonance and with $P_1 \leq 100$ days, only in 18 cases have the masses of both bodies have been measured or estimated with some accuracy. Of these, only 10 have orbital period ratios in the interval $P_2/P_1 \in [2.0, 2.10]$, and may be thus cataloged as members of the (near)-resonant region. This number continues to decline as we note that 7 planetary pairs have at least one of its members with $m_i > 50m_\oplus$, more than sufficient to open a gap in the disk and to have

Table 1. Mass measurements and orbital periods for Kepler-25 b,c and K2-24 b,c.

System	m_1 $[m_\oplus]$	m_2 $[m_\oplus]$	P_1 [d]	P_2/P_1	$m_0[M_\odot]$
Kepler-25	9.6 ± 4.2	24.6 ± 5.7	6.24	2.0390	1.22 ± 0.06
K2-24	21.0 ± 5.4	27.0 ± 7.0	20.89	2.0284	1.12 ± 0.05

migrated following a Type-II scenario. Since our model is based on analytical prescriptions for laminar Type-I migration, these systems are beyond the scope of our work.

Of the three remaining candidates, HD 219134 have recently been questioned. The first reference to this system appears in [24], who analyzed a total of 98 nightly averaged RV observations obtained with HARPS-N and found evidence of 4 low-mass planets. The outer member of the (near)-resonant pair was not detected. Later, [25] analyzed a total of 276 RV Doppler measurements and identified a total of 6 planets in this system. [26] found a substantial periodicity in the RV data due to stellar rotation with a period of 22.8 days, a value practically equal to $P_1/2$. Although the authors do not believe there is sufficient evidence to rule out the existence of the inner planet, the amplitude generated by the planet in the RV signal may be affected by stellar rotation and thus, the mass deduced for m_1 could in fact be substantially lower.

This leaves us with two systems, Kepler-25 and K2-24. Table 1 gives the masses and orbital periods of both systems, together with the stellar masses and the respective standard deviations. Both systems have nominal mass ratios larger than unity (i.e. $m_2/m_1 > 1$) and are thus candidates for resonant trapping in the 2/1 commensurability. We therefore proceeded to analyze whether the observed value of the resonance offset $\Delta_{2/1}$ could be achieved using our model with the assumption of a laminar flared disk.

Since the value of $\Delta_{2/1}$ is known, we can invert expression (4) to obtain explicitly the value of H_0 as function of the masses and the disk flare:

$$H_0^2 = \frac{2}{D\Delta_{\rm obs}^2}\left(C_1\frac{m_2}{m_0}\right)^2 \frac{\left[(1 - D(1+\beta))\mathcal{K}_2^*\left(C_2\frac{m_1}{m_2}\right)^2 + (B + D\beta)\mathcal{K}_1^*\left(\frac{\tau_{a_2}}{\tau_{a_1}}\right)\right]}{1 - \left(\frac{\tau_{a_2}}{\tau_{a_1}}\right)}, \tag{6}$$

where $\Delta_{\rm obs}$ is the observed value of the offset and $\mathcal{K}_i^* = 0.78(Q_a/Q_e)a_i^{-2f}$ are H_0-normalized expressions for the K-factors of each planet. If the planetary masses are known with even some accuracy, it is then possible to estimate relations between f and H_0 leading to the observed values of the offset. Since uncertainties in these values may be significant, we chose a statistical Monte Carlo approach incorporating the errors in the masses into the calculation.

For each system we ran a series of 1000 sets of (m_1, m_2) from a normal distribution with mean and variance as depicted from Table 1. From the values of each run, we then determined the distribution of values of (H_0, f) according to (6), for fixed values of α. Results are shown in color scales in Figure 2, where the top (bottom) panels correspond to Kepler-25 (KE-24), respectively. Left plots were drawn assuming $\alpha = 0.5$ while the right graphs are the results obtained considering $\alpha = 1.5$. As can be noted, outcomes appear only weakly dependent on the surface density profile, such that we have plotted only the extreme results, and not for $\alpha = 1$, as it is the interpolated plot between $\alpha = 0.5$ and $\alpha = 1.5$. The color code corresponds to the possibility of a pair (H_0, f) to be a solution of equation (6). Blue colors mean low possibilities while red is associated to higher frequency in the outcomes.

For Kepler-25, most of the positive results are occur along a broad curve with low values of H_0 and large values of the disk flare f. This is consistent with the Monte-Carlo simulations presented in [1] for the overall near-resonant population and systems with under-determined masses. Notice that the large uncertainty in m_1 does not significantly affect the result and the loci of values of (H_0, f) consistent with the observed offset remains fairly restricted.

Results for K2-24 appear less defined which implies that a wide range of disk parameters would lead to the observed resonance offset. In part this is due to the lower, and more easily achieved, value of $\Delta_{2/1}$

IOP Conf. Series: Journal of Physics: Conf. Series **911** (2017) 012027 doi:10.1088/1742-6596/911/1/012027

but also to particular planetary masses. On one hand, the individual masses are larger than for Kepler-25 which in itself leads to a wider resonance domain. On the other hand, the ratio of m_2/m_1 is almost unity which, as seen from equation (4) also implies larger offset even for low flare values and/or large H_0.

4. Conclusions

We present an application of the [1] model for resonance offset to two planetary systems (Kepler-25 and K2-24) close to a 2/1 MMR with significant observed offset and fairly established planetary masses. We find that the disk parameters necessary to explain the deviation from exact resonance under a laminar type-I migration are similar to those predicted by [1], namely a low disk scale-height and significant flare index. This agreement indicates that the proposed mechanism could indeed have played a dominant role in determining the observed distribution of (near)-resonant exoplanets.

Acknowledgments

This work has been supported by research grants from ANCyT, CONICET and Secyt-UNC. We are grateful to IATE and CCAD (UNC) for extensive use of computational facilities.

References

[1] Ramos X S e a 2017 *A&A* **602** A101 (*Preprint* 1704.06459)
[2] Marcy G W e a 2014 *ApJ S* **210** 20 (*Preprint* 1401.4195)
[3] Petigura E A e a 2016 *ApJ* **818** 36 (*Preprint* 1511.04497)
[4] Hansen B M S and Murray N 2012 *ApJ* **751** 158 (*Preprint* 1105.2050)
[5] Petrovich C, Malhotra R and Tremaine S 2013 *ApJ* **770** 24 (*Preprint* 1211.5603)
[6] Chatterjee S and Tan J C 2014 *ApJ* **780** 53 (*Preprint* 1306.0576)
[7] Rein H 2012 *MNRAS* **427** L21–4 (*Preprint* 1208.3583)
[8] Baruteau C and Papaloizou J C B 2013 *ApJ* **778** 7 (*Preprint* 1301.0779)
[9] Paardekooper S J, Rein H and Kley W 2013 *MNRAS* **434** 3018–29 (*Preprint* 1304.4762)
[10] Quillen A C, Bodman E and Moore A 2013 *MNRAS* **435** 2256–67 (*Preprint* 1304.6124)
[11] Beaugé C, Michtchenko T A and Ferraz-Mello S 2006 *MNRAS* **365** 1160–70 (*Preprint* astro-ph/0404166)
[12] Michtchenko T A, Beaugé C and Ferraz-Mello S 2008 *MNRAS* **391** 215–27
[13] Beaugé C, Ferraz-Mello S and Michtchenko T A 2003 *ApJ* **593** 1124–33 (*Preprint* astro-ph/0210577)
[14] Ramos X S, Correa-Otto J A and Beaugé C 2015 *CMDA* **123** 453–79 (*Preprint* 1509.03607)
[15] Ferraz-Mello S 1988 *AJ* **96** 400–408
[16] Lee M H 2004 *ApJ* **611** 517–527
[17] Lee M H and Peale S J 2002 *ApJ* **567** 596–609
[18] Cresswell P and Nelson R P 2008 *A&A* **482** 677–90 (*Preprint* 0811.4322)
[19] Papaloizou J C B and Larwood J D 2000 *MNRAS* **315** 823–833
[20] Tanaka H and Ward W R 2004 *ApJ* **602** 388–95
[21] Tanaka H, Takeuchi T and Ward W R 2002 *ApJ* **565** 1257–74
[22] Cresswell P and Nelson R P 2006 *A&A* **450** 833–53
[23] Goldreich P and Schlichting H E 2014 *AJ* **147** 32 (*Preprint* 1308.4688)
[24] Motalebi F e a 2015 *A&A* **584** A72 (*Preprint* 1507.08532)
[25] Vogt S S e a 2015 *ApJ* **814** 12 (*Preprint* 1509.07912)
[26] Johnson e a 2016 *ArXiv e-prints* (*Preprint* 1602.05200)

Institute of Physics
Dirac House, Temple Back
Bristol BS1 6BE UK

ISSN: 1742-6588
ISBN 978-1-5108-5039-2

71st Annual Meeting of the Labor and Employment Relations Association and LERA@ASSA Meeting 2019

Labor and Employment Relations Association Series

Atlanta, Georgia, USA
4-6 January 2019

Cleveland, Ohio, USA
13-16 June 2019

LERA